AF478209

Innovation, Science, Environment

Special Edition:
Charting Sustainable Development
in Canada, 1987–2007

Edited by
GLEN TONER and
JAMES MEADOWCROFT

Published for
The School of Public Policy and Administration
Carleton University
by
McGill-Queen's University Press
Montreal & Kingston · London · Ithaca

Legal deposit second quarter 2009
Bibliothèque nationale du Québec

Printed in Canada on acid-free paper that is 100% ancient forest free
(100% post-consumer recycled), processed chlorine free

McGill-Queen's University Press acknowledges the support of the Canada
Council for the Arts for our publishing program. We also acknowledge
the financial support of the Government of Canada through the Book Publishing
Industry Development Program (BPIDP) for our publishing activities.

Library and Archives Canada has catalogued this publication as follows:
Innovation, science, environment : Canadian policies and performance,
Special edition: charting sustainable development in Canada, 1987–2007
/ edited by Glen Toner and James Meadowcroft.

(Innovation, science, environment series ; 4)
ISBN 978-0-7735-3532-9 (bound). – ISBN 978-0-7735-3533-6 (pbk.)

1. Environmental policy – Canada – Evaluation. 2. Sustainable
development – Government policy – Canada. 3. Technological
innovations – Government policy – Canada. 4. Science and state – Canada.
I. Toner, Glen, 1952– II. Carleton University. School of Public Policy
and Administration III. Series.

GE190.C3I55 2009   352.7'450971   C2009-900415-9

This book was typeset by Interscript in 10/12 Minion.

Glen Toner dedicates this volume to Professor Gurston Dacks, his mentor and MA thesis supervisor in the Department of Political Science, University of Alberta, 1976–78.

James Meadowcroft dedicates this volume to William Lafferty, whose own scholarship in the field of sustainable development has always been an inspiration.

Both editors further dedicate this volume to Charles Caccia and Jim MacNeill, Canadian pioneers in the sustainable development revolution.

Contents

Preface

This is the fourth edition of our annual volume of commentary and assessment of Canadian and related comparative innovation, science and environment (ISE) policies and institutions. It emerged out of a broader body of research and teaching at the Carleton Research Unit on Innovation, Science and Environment (CRUISE) and the School of Public Policy and Administration at Carleton University.

This year's volume is the first to have a single theme: *Charting Sustainable Development in Canada 1987–2027*. The goal is both to celebrate the 20th anniversary of the publication of *Our Common Future*, the 1987 Report of the World Commission on Environment and Development, and to use the occasion to reflect on the progress which Canada has made in engaging with sustainable development over the past two decades and to consider the perspectives for the future. The contributions that make up this volume come both from 'old hands', scholars and practitioner who have been concerned with sustainability for some time, and from some new voices. In each case we hope that the authors' efforts will stimulate the reader's own reflection on the important challenges ahead.

Thanks are due to Sheena Kennedy and Kimmie Huang in the School of Public Policy and Administration for their excellent research and technical support, and to Joan McGilvray and her colleagues at McGill-Queen's University Press for their always professional editorial and publishing expertise. David Runnall's staff at the IISD were instrumental in organizing the conference which seeded the majority of these papers. These publications flow from the continuing support and scholarly stimulation provided by our colleagues at CRUISE and in the School of Public Policy and Administration at Carleton University.

Glen Toner and James Meadowcroft
Ottawa
January 2009

1 Engaging with Sustainable Development: Setting the Canadian Experience in Context

JAMES MEADOWCROFT
AND GLEN TONER

This volume assesses the progress Canada has made in taking up the challenge of sustainable development and perspectives for the future. More than twenty years have now passed since the World Commission on Environment and Development (WCED) published *Our Common Future*, launching the idea of sustainable development onto the world stage.[1] Over this period the idiom of sustainability has become embedded in Canada's political vocabulary. Sustainable development is mentioned in federal legislation and features in the mandates of several departments. The country has a Parliamentary Commissioner for the Environment and Sustainable Development and in 2008 Parliament adopted a Federal Sustainable Development Act. But what has really changed? To what extent has Canada begun to come to grips with environment and development dilemmas? And what are the obstacles to moving forward? These are the issues this collection of essays seeks to address.

Most of the chapters first took form as presentations to the conference "Facing Forward, Looking Back: Charting Sustainable Development in Canada 1987–2007–2027," held at the National Arts Center in Ottawa in October 2007. The meeting was organized by the International Institute for Sustainable Development (IISD), the Carleton Research Unit in Innovation, Science and Environment (CRUISE), and the Canada Research Chair in Governance for Sustainable Development at Carleton. These papers were revised subsequently by their authors and, together with a number of additional commissioned chapters, are presented in this special issue of *Innovation, Science and the Environment*.

Sustainable development is a broad topic, so it is important to establish at the outset that the centre of gravity of this volume is governance, particularly at the federal level. It does not have chapters on individual environmental themes (water, climate, forests and so on). Nor does it deal at length with provinces and municipalities. And the issue of Canada's contributions at the international level is only touched upon. This is not because these issues are unimportant, but because everything could not be included in a single volume. What the essays published here do deliver is a broad overview of engagement with sustainable development in Canada over the past two decades, and of the important challenges for the future.

INTERNATIONAL ENGAGEMENT WITH
SUSTAINABLE DEVELOPMENT

Although arguments about sustainability had been underway in international environment and development circles for some time, it was the publication of *Our Common Future* in 1987 that brought sustainable development to broad international attention. In the now famous words of the WCED, sustainable development was defined as "Development that meets the needs of the present without compromising the ability of future generations to meet their own needs."[2] And they went on to add that "It contains within it two key concepts: the concept of needs, in particular the essential needs of the world's poor, to which overriding priority should be given; and, the idea of limitations imposed by the state of technology and social organization on the environment's ability to meet present and future needs."[3] At the United Nations Conference on Environment and Development, also known as the Earth Summit, held in Rio in 1992, leaders from around the world officially endorsed sustainable development as an international goal. And in the following years sustainable development was gradually incorporated into the language of international governance.[4]

A central argument in *Our Common Future* was that something had gone profoundly wrong with international development. For, on the one hand, advance remained stalled in many developing countries, where millions still languished in poverty. And, on the other hand, rapidly expanding environmental problems threatened to undercut the foundations on which social prosperity was based. Underlying the argument was a realization that it would be impossible to extend the profligate consumption habits of the prosperous North to peoples around the world without imposing a grave strain on global ecosystems. Hence the call to change the "quality of growth," placing emphasis on meeting the basic needs of those who had least, while simultaneously striving to reduce dramatically the burden placed on the global environment.

In some ways the world of 1987 was very different from the one we know today. Europe remained divided between East and West, and the rivalry of the two superpowers – the USA and the USSR – dominated international affairs. Apartheid still prevailed in South Africa. China's rapid economic rise remained in the future. Personal computers, mobile phones and the internet were in their early days. And climate change was only beginning to emerge on to the international political agenda. Yet the critical tensions between environment and development were already clear. So *Our Common Future* emphasised that if progress and prosperity were to lie ahead, a transformation of the international development trajectory was essential.

Sustainable development is best understood as a new international norm that can guide environment and development decision making.[5] It refers to a basket of normative ideas including concern with human wellbeing, the protection of the environment, equity between and within generations, and participation in environment and development decision-making. Like other value laden concepts – such as justice, freedom and democracy – it is subject to continual debate and re-interpretation. And yet, also like these concepts, it has an underlying meaning that helps structure debate. One cannot "read off" the correct course of action in any specific situation directly from the concept of sustainable development. Analysis and argument are required in every case to determine an appropriate way forward. But sustainable development provides a reference point and common idiom with which to carry forward this enterprise.

The natural environment underpins societal development: it provides resources for economic activity, ecosystem services to maintain life, amenities for recreation and enjoyment, and a focus for aesthetic and ethical concern. Yet human activities increasingly threaten this foundation. Year after year scientific studies detail the rising pressures on the natural world imposed by our technologies, consumption patterns, and population growth.[6] Climate change is perhaps the most obvious example. Modern industrial and agricultural practices – especially the combustion of fossil fuels which have powered economic advance since the Industrial Revolution – are changing the planet's climate.[7] And this shift presents serious risks for future human generations, to say nothing of the welfare of the earth's other inhabitants. But there are many other areas where we are imposing critical strains on global ecosystems, including the nitrogen cycle, patterns of land and water use, chemical releases, and the unsustainable exploitation of biological resources. The methodical destruction of the world's fisheries, which continues unabated, shows the difficulty of taking action even in the face of imminent threats.

To move away from current practices and to adopt a more sustainable development orientation requires some pretty radical changes. Economic activity must be "decoupled" from increased environmental loading, and pressures must be brought back within the limits of natural ecosystems. Scientific models suggest that, if we are to avoid some of the more serious climate risks, global greenhouse gas emissions will have to peak in the next decade, and cuts of 80% or more on 1990 levels will be required by mid century by developed states. This implies fundamental change in key economic sectors including energy, transport, construction, and agriculture to develop a low carbon economic future. Increased resource efficiencies, and new technologies are vital. But over time changes will also be necessary in areas we are less comfortable about discussing, including levels of population and consumption. Environmental impacts are in part determined by just how many of us there are, and sooner of later this issue will return to haunt international climate negotiations. But for those of us living in rich countries like Canada consumption may seem to be the more pressing issue. There is little doubt that our current patterns of consumption are seriously undermining ecological integrity. Twenty years ago the authors of *Our Common Future* noted that "Perceived needs are socially and culturally determined, and sustainable development requires promotion of values that encourage consumption standards that are within the bounds of the ecologically possible and to which all can reasonably aspire.[8] Here again natural limits (the ecologically possible) and equity (to which all can reasonably aspire) are to provide the framework for re-orienting development pathways.

At the core of the notion of sustainable development is the idea of integrating environment and development decision making. The goal is to "factor-in" the environmental dimension from the outset. Rather than first determining a development orientation, and later having to enact environmental policy (to clean up the mess and forestall further deterioration), the point is to consider environmental implications from the outset. Thus the management of the environment is to be understood as part of what counts as authentic development. Integrating environment and economic decision making has important implications for political and economic actors. On the economic side the true cost of environmental goods and services needs to be integrated into economic calculus, and a duty of environmental care must become part of everyday business practices. On the government side environmental issues need to become an explicit focus across public administration, and the environmental consequences of public policy should be consistently examined.

Over the past century developed countries have proven relatively successful in engineering economic growth. And while economies go through ups and downs, the material conditions of our existence are vastly superior to those enjoyed by citizens a century ago. In terms of health and social policy, the provision of education and welfare services, we have also done quite well, although there are still gaps and inequities in our institutions. But on the environment front we have not done so well. Since the creation of the institutions of modern environmental governance in the early 1970s (departments, agencies, advisory bodies, laws, and so on), emissions of many major pollutants have declined, contaminated sites have been cleaned up, and protected areas have increased. Environmental policy has delivered benefits, and with the turn towards sustainable development there has been more of a focus on pollution prevention and on transforming production to avoid environmental harm. At the level of government, companies, and community action, the environment is more present than ever before. And yet, *despite all this*, efficiency gains continue to be swamped by the rise in absolute consumption and material throughput.[9] On a global level environmental pressures are continuing to accumulate. Indeed, the prosperity we enjoy has been based on the degradation of global ecosystems.

Confronting these problems requires institutional innovation. It requires reforms to political, economic and administrative organizations that were designed in earlier times to solve problems of earlier generations. But transforming institutions is not an easy task. And it is here that the challenge lies in passing from talk about sustainable development to action.

HOW ARE OTHER COUNTRIES DOING?

As we reflect on Canada's engagement with sustainable development it is worth pausing briefly to set these efforts in context by considering the experience of other highly developed states. Of course, national circumstances differ enormously. Ecological endowments, industrial structures, political institutions, and administrative cultures vary from country to country. There are still relatively few detailed cross-national studies of relative performance. And one must in any case be careful about conclusions drawn from international comparisons. Nevertheless some basic observations are possible.

Over the past twenty years developed countries have introduced more stringent environmental laws, elaborated integrated approaches to environmental management, diversified the range of policy instruments, devoted increased attention to international cooperation, and taken some steps to institutionalize sustainable development.[10] Areas on which there has been movement include: tightening controls on air and water pollution, strengthening chemical management systems, phasing out ozone depleting substances, extending environmental monitoring and reporting systems, increasing waste recycling, and enlarging protected areas.[11] Some countries have introduced binding greenhouse gas emissions controls and have stabilized or begun to reduce these emissions. New renewable energy sources, especially wind, have started to come on stream. And many businesses are actively engaged with the environment and sustainable development.

Yet overall, experience is uneven   varying from sector to sector, from problem to problem, from country to country, and over time. A comparatively small number of states stand out for the enthusiasm with which they have engaged with issues around the environment and sustainable development. Sweden has been perhaps the most consistent international leader, but at various points and on various issues the Netherlands, Denmark, Finland,

and Germany have made a mark.[12] Even the United Kingdom, which in the mid 1980s had been dismissed as "the dirty man of Europe" has played a positive role, particularly in relation to climate change.[13] The contribution of such states in part accounts for why the EU as a whole has been seen as a relatively positive force on environment and sustainable development issues over the past decade. But there are also other factors at play.

During the 1990s the European Commission took on a more active role in environmental affairs, leading to a partial harmonization of environmental law across the EU.[14] And the expansion of the EU has meant a transfer of the older member states" environmental "acquis communautaire" to the new members, generally improving their environmental standards. And, as the EU has looked to play a more active role in international affairs, issues like climate change seemed well suited to its "soft power" resources. Nevertheless, it should not be forgotten that the absence of international leadership on climate change from other quarters (such as the United States, Australia, and Canada, all of which had problems in relation to the Kyoto Protocol), has made the EU performance look good by comparison. Of course, the EU is no sustainable development ideal. Positions of individual countries vary substantially. Some new member states, such as Poland, are leading resistance to more stringent climate measures, and it is far from clear that the EU's international leadership role on climate change will continue.

In terms of institutional innovations related to sustainable development one of the most obvious changes has been the introduction of national plans and strategies. Almost all developed states now have such an instrument, but their status differs considerably.[15] In some countries these are simply cosmetic documents, gathering together lists of pre-existing government activities, which have been prepared for reporting to international meetings. Elsewhere strategies that initially had some political salience have been allowed to lapse. But in a number of countries these strategies have been taken seriously. Although they are not operational plans, they are playing an important communicative role (within government, between government and society), and helping to establish priorities and to orient societal actors for the changes sustainable development entails. If strategies can become embedded in institutionalised cycles of planning, implementation and review they can contribute to political and administrative engagement with sustainable development.[16] Recently, the EU has introduced a process of international "peer review" of such strategies.

Sweden has already been cited as a consistent leader in the area of environment and sustainability and it is worth noting some of the innovations introduced there in recent years. In the early 1990s Sweden underwent a serious financial crisis which called into question its traditional welfare state model. Rather than abandoning the environment and sustainable development in the face of these economic difficulties, the Labour Party issued a call for building a "green welfare state," adjusting the established image of Sweden as the "Peoples Home" to that of a "Green Peoples Home."[17] The government introduced spending programs for green job creation. As a political move this helped: a) further entrench environment and sustainable development as part of the Swedish national identity; and, b) bind together economic, social with environmental goals. Another important innovation was the introduction of a comprehensive Environmental Code which systematized regulation. Sweden also has become the first developed country to call officially for the phasing out of petroleum dependence by 2020.

Perhaps most impressive of all has been a system of national Environmental Objectives. There are 16 general objectives adopted by Parliament that cover the whole environmental domain. Through a protracted process of consultation with stakeholders and the public, these objectives were operationalized into specific quantitative goals, with clear timelines, and interim targets. The contribution from each county and municipality in meeting the national goals was established, and specific ministries and agencies made responsible for progress towards each objective. Performance is monitored by a National Objectives Council, made up of important stakeholders.[18] The original – extremely ambitious – goal of this program was to resolve all of Sweden's domestic environmental problems within a generation. The detailed reports issued by the National Objectives Council make clear that the country will not attain this goal by 2020. But this impressive system of "management by objectives"[19] represents perhaps the most comprehensive approach to managing environmental burdens attempted anywhere.

Of course, Sweden, with a unitary state and relatively small and homogenous population is very different from Canada. And one sometimes hears that it is a mistake to point to such an exceptional case. But there are other countries which also reveal interesting patterns of innovation. Germany, the largest of the EU states, with the most developed industrial base, has made considerable headway on many issues. In the area of renewable energy, for example, Germany pointed the way with its adoption of "feed in tariffs" (that oblige utilities to purchase renewable electricity at fixed rates for a specified period). Germany now has more than 22 Gigawatts of installed wind capacity. And in recent years photovoltaics and cogeneration are also accelerating strongly. Related to the growth in renewable energy has been the expansion of green employment, export markets, and the green business sector. The United Kingdom has also become something of a leader on the climate front. Like Germany, it benefited from fortuitous circumstances that helped get its greenhouse gas emissions on a downward track (in Germany the collapse of the East and closure of old plants, in the UK the switch from coal to gas for electricity generation). But as in Germany, policy initiatives account for additional reductions. Innovations in the UK include establishment of a range of specialized institutions working towards a low carbon economy (such as the

Carbon Trust). Of particular interest is the new Climate Change Act that establishes a system of national Carbon Budgets that must be approved by Parliament.[20] The Act also creates a Climate Change Committee with a statutory duty to assess the Carbon Budgets and determine whether they are on track to meet the long term target of 80% emissions reduction on 1990 levels by 2050. What is remarkable about this arrangement is that it institutionalizes the idea of a carbon constrained world, establishing that the country must keep within its carbon budget just as it must manage its financial resources.

As these examples show, there is a lot going on out there. Of course, each of these countries also has its problems. Even those who have taken up this challenge most consistently have still not managed to curtail their overall environmental footprint, and shift their development decisively onto sustainable lines. Yet in many cases the accomplishments are encouraging.

Although Canada ranks quite highly in certain international rankings (in 2008, 3rd on the UN Human Development Index[21] and 12th on the Yale Environmental Performance Index),[22] Canada does not score so well on comparisons that focus on environmental policy and regulatory standards. Compared to other OECD countries, for example, Canada places near the bottom on many environmental indicators.[23] And despite a promising beginning, on sustainable development Canada now lags significantly behind the most innovative jurisdictions. Many elements account for this weakness. Perhaps most obvious is the relative abundance of Canada's environmental endowments. Canada has such a generous resource base – a large territory with a low population density, rich forests and agricultural lands, plenty of water and energy resources of every kind (oil, gas, coal, uranium, hydro power, and so on) – that it is sometimes hard to appreciate natural limits. On climate change worries about moving more rapidly than the principal trading partner have played a role, as has the lobbying of special interests, and the ideological influence of climate sceptics transmitted through the US media. Much is usually also made of Canada's decentralized federation and peculiar constitutional arrangements that vest substantial powers in sub-national units. Clearly this often makes decisive government action more difficult, but broader political issues and the beliefs and priorities of Canadian voters must to a large extent be held accountable.

IS THE PROBLEM SUSTAINABLE DEVELOPMENT ITSELF?

Over the years sustainable development has been subject to a variety of criticisms. At the broadest level some environmentalists have complained that sustainable development puts too much emphasis on development and not enough on the environment. They argue that it is the environment that should be sustained by sustainability, not some human-centred process of "development." Environmental destruction has always been justified in the name of growth, and "sustainable" development promises more of the same. What is needed is not a new kind of development, but more consistent environmental concern. In other words, sustainable development is bad for the environment. On the other hand, different voices (especially in the developing world), have suggested that sustainable development is little more than a ploy by environmentalists (and others in the rich countries), to limit the poorer countries" quest for affluence. They argue that throughout the twentieth century the rich countries did not hesitate to sacrifice their environments to achieve economic advance, and today developing counties are entitled to make sovereign decisions about the use of their own natural resources. So here the accusation is that sustainable development is *too* environmental.

That sustainable development has been subject to both sorts of critique reflects its character as a bridging concept explicitly formulated to try to draw together environment and development decision making. It is perhaps not surprising that those who are more or less exclusively concerned with one element or the other are dissatisfied. Nevertheless, it is worth considering these critiques further, especially in a context where real progress in transforming the global development trajectory over the past two decades has been quite modest. After all, a sceptic could say: with only limited results from all this effort, perhaps we should revisit the possibility that there is something fundamentally wrong with the idea of sustainable development itself.

Turning first to the environmental critique, the charge is that sustainable development is concerned with sustaining development rather than with sustaining the natural environment. In one sense the accusation is entirely true, but in another it misses the point. Certainly sustainable development is about development. Just look at the way the expression is actually put together: "sustainable *development*" – clearly it is the development that is to be sustained. But what is development? Development is understood as a process of human advance. It relates to economic, social, political and cultural change. It involves science, education, health, the arts and civic life. Above all, as *Our Common Future* emphasised, development should not be reduced to economic growth, although appropriate economic growth may be an important element of development. Sustainable development is development that does not undermine its own preconditions, taking care of the future as well as the present. It is development that pays particular attention to environmental limits and to the satisfaction of basic human needs. To this extent it is authentic development, which stands in contrast to phoney, misguided or unsustainable "development" that may increase immediate economic returns, but which does so at the cost of other values. Sustainable development emphasises that the environment is a critical determinant of human well being. Unless ecosystems, biodiversity, and environmental quality are maintained, sustainable development will be impossible. Thus environmental sustainability is a critical dimension of sustainable development.

But major environmental problems can never be solved by focusing on the environmental dimension alone. Of course, a specific incidence of pollution, or a distinct threat to a natural habitat, can be managed through specific environmental intervention. But the major environmental problems we face are related to large scale economic and social realities. And progress can only be made by considering the economic and social matrix in which they are embedded. In other words, the quest for sustaining environments independent of the effort to redefine the societal development trajectory is ultimately hopeless. After all, the destruction of the natural environment is typically the by-product of other (largely economic) activities. And these activities have to be reconsidered if there is to be real movement on the environmental front. Hence the insistence on drawing together environmental, economic and social decision making.

Does all this mean that sustainable development is hopelessly anthropocentric – capable of relating only to human ends and human goods? Certainly sustainable development starts from the human perspective, because it is human social development to which it refers. But things do not have to stop there. In the first place, the report of the WCED argues that there are non instrumental reasons for valuing nature, noting that "the conservation of nature" is "part of our moral obligation to other living beings and future generations."[24] More importantly, since sustainable development is about defining an appropriate development trajectory, and building the kind of world we want to leave to posterity, it is open to environmentalists to argue that a more substantial ethical recognition of non human nature is a critical element of what true development entails. After all, moral advance is just as much part of a sustainable development trajectory as other dimensions of human experience. So rather than closing down the debate, sustainable development opens it up to encourage reflection over the true nature of human societal progress.

Turning next to the developmental critique, the charge is that sustainable development represents an imposition of "Northern" environmental values on the developing world. There is a suggestion that, having climbed the development ladder, the rich countries now want to change the rules to make it harder for those following behind. And this is typically coupled with another claim – that environmental concerns arise naturally, and can be addressed appropriately, at an advanced stage of economic development. Once a country has industrialized, met the basic needs of its populations, and entered the ranks of the high (or upper middle) income countries, then it will have the inclination and resources to tackle environmental issues. In other words, first let developing countries get rich, and then they can worry about the environment. On this line of reasoning the sustainability in sustainable development is simply a distraction for many countries.

The first thing that needs to be said in response to this argument is that developing countries are not a homogeneous group. In fact, they are characterised by huge differences in wealth, size, population and living conditions, with some countries like South Korea approaching the situation of the developed states, others like Brazil and China growing rapidly, and many more nations still mired in extreme poverty. Development and environment priorities will differ markedly according to particular circumstances. Countries lacking functioning governments, with stagnant economies and large scale privation, will focus on building basic institutions, growing their economies, and meeting basic needs.

Yet it is also clear that developing countries cannot afford simply to forget about the environment until their economic goals have been met. Efforts to meet basic needs are often entwined with environmental issues, such as securing fresh water, fuel, and agricultural livelihoods. Environmentally deleterious practices (such as forest clearances) have negative long term economic impacts as well as serious public health implications. The conditions in which today's developing countries find themselves are very different from those that confronted the now rich countries a century ago. Absolute population levels and population growth rates of today's developing countries are generally higher. The technologies and industrial processes now available are more powerful, but also vastly more environmentally destructive. The natural environments are themselves quite different, and often more vulnerable. And of course there are now global environmental problems that had not previously emerged.

This does not obviate the responsibility of rich countries to move decisively on issues such as climate change, to reduce their own emissions and to provide financial and technical assistance to poorer states to green their development pathways. But it does mean that large and rapidly growing developing countries will be expected also to contribute to long term emissions reductions. Because without these efforts it will be impossible to avoid serious climate risks.[25]

Another argument that crops up from time to time goes something like this: sustainable development is a stupid idea because it brings normative claims (such as equity and participation) into discussions about the management of specific environmental problems. Sustainable development emphasises interconnectivity, and suggests that we should integrate all sorts of economic, social and environmental factors in decision making. Of course the world is complex, and everything is connected to everything else. But the way we solve real world problems is by focusing on specific issues one by one, not by lumping everything in one basket. By suggesting that we have to solve economic issues, social and equity issues and environmental issues all together, sustainable development actually means we will solve none of these problems. This line of attack might be called the engineering critique of sustainable development. Engineers typically work by breaking problems down to specific manageable subsystems and then reassembling the parts to generate an efficient whole. And from this perspective sustainable development appears like a complicated mish-mash that stands in the way of progress.

The trouble with this is that it ignores much of the reality of political and social systems. Environmental problems, such as climate change, cannot be stripped out from the societal context within which they are embedded. The problems arise from practices that are rooted in economic, social and political structures. Environmental problems always involve equity issues, concerning who bears the costs of environmental degradation and who pays the costs of remedial measures. Sometimes these dimensions are not explicitly addressed in the policy process, but typically this means that the voices of certain groups are simply being silenced. And on the big issues, at least in democratic countries, there is no way to avoid fuller debate. For example, any economy wide measure designed to deal with carbon emissions will have equity implications. Sustainable development suggests that it is better to consider these issues up front. And, because it is impossible for a centralized administration to appreciate all possible impacts of a policy response; because bad environmental practices have often been nurtured by secrecy with the supposed experts calling the shots; and because citizens are entitled to influence decisions about development that impact upon their lives; sustainable development emphasises public and stakeholder participation in decision making.

Of course, beyond all this, there may still be those who feel that the term "sustainable development" has been compromised by the way certain politicians and corporate leaders use it to justify "business as usual." But all political concepts are subject to such co-optation How many mis deeds have we heard justified in the name of freedom and democracy. But that does not mean that we should abandon these essential values but rather struggle to affirm their real implications.

Sustainable development captures a critical dimension of the modern world. It expresses the need to radically transform the existing development trajectory. And were we to abandon the concept we would simply need to coin another word to express this normative orientation. So if this is true, and the problem is not the idea of sustainable development itself, why is progress so slow? Why have environmental pressures continued to accumulate despite all the conferences and roundtables, the government laws and programs, and the actions of activists and good corporate citizens?

Much could be written on this topic, but here it is sufficient to note that the transformation suggested by sustainable development really is ambitious. Shifting the orientation of societal development is not easy to accomplish. It rivals some of the most ambitious changes in human history, and requires reforms to political and economic institutions at all levels. Just how deeply these reforms need to go cannot be determined with certainty in advance. All that can be done is to change what appears to block progress, assess the results, and then dig deeper as necessary. Major changes of this type typically take time to work their way through. This is not to suggest that things always move at a glacial pace and that only incremental change is possible. Things can speed up. Crises and revolutions punctuate human history, sweeping away established ways of doing things almost overnight. But such rapid bursts may be preceded by many decades of apparent inertia where slow change matured under the surface.

On the side of "structure," established institutions continue to frustrate innovation, "locking in" traditional ways of doing things that have lost whatever societal rational they may once have possessed. More "sustainable" technologies, business models and administrative practices face countless obstacles related to ingrained habits, standard operating rules, codes and standards, regulatory regimes, access to resources, educational practices, and so on.

On the side of "agency," powerful interests are at work, attempting actively to block change. Many influential groups and individuals believe they have done well from existing economic and political arrangements. They are happy for things to go on more or less as they are, and they are prepared to mobilize power resources in the attempt to delay change indefinitely. And, of course, there is also the role of ideas – where outmoded conceptions of resource intensive economic growth, old-fashioned business practices, and bankrupt political and administrative approaches remain current. All this is not to suggest that change is impossible, but rather to emphasise that it will not come easily. Building coalitions for reform and achieving change on the desired scale will be the task of decades.

CONTENTS OF THIS VOLUME

The contributions that make up this volume offer a variety of insights into Canadian engagement with sustainable development and perspectives on the future.

The first piece is the speech which DAVID RUNNALLS presented in the fall of 2007 to the conference "Facing Forward Looking Back" held to mark the twentieth anniversary of the publication of *Our Common Future*. David Runnalls has been involved with the sustainable development agenda practically from the beginning, and as president of the International Institute for Sustainable Development he has done much to champion the cause of sustainability, both in Canada and internationally. We therefore thought it important to share with our readers his perspectives on the development of activity around this issue. His speech highlights the transformative agenda associated with sustainable development. It stresses the innovative role played by Canada in the early years, while also noting factors that contributed to a later dropping away of interest. It emphasizes the magnitude of the challenges that still confront Canada, while remaining positive about the possibilities of change.

In their chapter, GLEN TONER and FRANÇOIS BREGHA assess the Canadian federal government's response since the late 1980s to the injunction in

Our Common Future to make institutional and policy changes to support the transformation to a sustainable future. Canada's record over this twenty year period has been mixed. A number of framework policies and institutional innovations were introduced but their impacts were often blunted at the implementation stage. Sustainable development capacity building at the societal, corporate, and governmental levels has grown but the concrete results have been disappointing. The federal policy response has emphasised voluntary and regulatory approaches to a greater extent than economic instruments. "Internal to government" instruments like strategic environmental assessment of cabinet decisions, greening of government operations, information programs, and sustainable development strategies have never been allowed to reach their full potential. The essay proposes a number of additional policy and institutional changes that are required to re-establish Canada's reputation as a world leader ... rather than a laggard, in governance for sustainable development.

ANN DALE discusses obstacles to a more meaningful engagement with sustainable development in Canada by established governance structures. She argues that governance mechanisms systematically fail to reflect societal dependence on the natural world, and talks of the "solitudes, silos and stovepipes" that act as barriers to the integrated thinking and action required to promote sustainable community development. Dale points to the importance of "interdisciplinarity," "transdisciplinarity," long term planning, and "place sensitive" approaches in order to promote movement. And she champions a "networked governance model" that can enhance policy coherence and address the multiple-dimensions of sustainability.

MARK WINFIELD argues that environmental policy in Canada has always been highly sensitive to the public salience of environmental issues. Canadian history suggests that four major waves of high public concern drive periods of major legislative, institutional, policy and program innovation. But these, have been followed by long periods of low public salience in which governments retrench policy activity and slow implementation. The association with cycles of environmental concern makes it difficult for sustainable development to be addressed on a routine basis. Winfield argues that stronger framework policies and institutions can help soften the cyclical swings. Thus, ecological fiscal reform, strengthened regulatory capability, enhanced science and education regimes, the embedding of policy commitments in international agreements, and more robust mechanisms for independent evaluation and public reporting would entrench sustainability issues in economic and social policy concerns as opposed to exclusively environment considerations.

ROGER GIBBINS reminds us that the Canadian federation is difficult to govern. A long term transition to a low carbon energy economy will impact differently across the regions of the country, and Canada has not been well served by a traditional party system that further exacerbates regional tensions as a result of the single member plurality electoral system. Gibbins argues that climate change policy is really about energy policy and asks "how do we construct a national energy strategy for a carbon constrained future ... without exacerbating regional tensions within the federation." He suggests that if climate change becomes so important that Canadians insist their governments take meaningful action then this could tilt the federal balance of power in favour of the national government. Given the implied threat of separation and alienation by certain regions if the response is shaped by redistributive objectives or regional insensitivity, Gibbins argues that we must take care to get the design of the new policy system "right."

Our Common Future stressed the intergenerational nature of the sd challenge. The knowledge, values, skills, and behaviours of humans are determinant of their impact on other species and the planet. DAVID V. J. BELL traces the influence that *Our Common Future* and its offspring like the Rio Earth Summit's *Agenda 21*[26] have had on international and Canadian efforts to introduce Education for Sustainable Development (ESD) into both the formal and informal educational systems. In building the case that cultural values are central to both policy and societal behaviour, Bell argues that it is not a question of whether to educate or use regulation or economic policy instruments. Both immediate and long term solutions and instruments are required. In fact, Canadian engagement with ESD has been episodic with spurts of enthusiasm mixed with periods of indifference. Bell draws on recent developments at the provincial level to suggest that Canada was again moving forward in 2008 with a broad range of activities emerging in part in response to international drivers such as the United Nations Decade of Education for Sustainable Development (UNDESD) 2005–2014.

DAVID WHEELER and ANNIKA TAMLYN argue that business and industry have a crucial role to play in achieving sd in Canada. They are concerned that several of Canada's most innovative corporate sustainability practitioners (Falconbridge/Noranda, Dofasco, Zenon Environmental, Alcan, and Shell Canada) were taken over in the last couple of years by non-Canadian firms. The authors compare corporate Canada's sustainability engagement and performance against nine other developed countries and conclude that Canada ranks sixth, though fourth within the G8. They go on to explore the proposition that small and medium sized enterprises may be sufficiently important to future sustainability challenges for Canada that we should pay more attention to them both from a research and public policy perspective. In addition to the traditional resource based sectors, an urgent challenge for Canada- and an opportunity- is for leaders in the high technology, services, retail, and manufacturing sectors to begin driving the sustainability agenda. This will require the leaders of key businesses in these sectors (think Loblaws and Research in Motion) to recognize their roles both as value-creators and social actors. For their part, governments will have to create the right conditions for success through their own procurement practices and regulatory activities, rewarding firms large and small with market opportunities and sustainability pay-offs.

The 1986 visit to Canada by the WCED had a very concrete outcome. The National Round Table on the Environment and Economy (NRTEE) can trace its roots to the synergies created by the National Task Force on the Environment and Economy during this period. SERENA BOUTROS explores a fascinating story of institution building which brought together a broad range of collaborators (including some traditional adversaries) in the industrial and environmental communities. The history of the NRTEE shows the powerful influence of key leaders as they managed and balanced the shift in emphasis amongst catalytic, advisory, and advocacy functions. As a small and independent agency with access to ministers and the PM it has had particular opportunities to advise government decision makers while simultaneously attempting to build networks of common interest amongst divergent sectors of society. The NRTEE celebrated its 20th anniversary in the fall of 2008. Boutros reflects on its achievements and setbacks in the face of a constantly changing political environment and increasingly formidable sustainability challenges. The prominent role played by the NRTEE on the climate change issue during the Harper Conservative regime is the most recent example.

One of Canada's most successful institutional innovations in this area is the International Institute for Sustainable Development (IISD). LILLIAN HAYWARD explores the evolution of the Winnipeg based Canadian success story. IISD was part of the Canadian response to work of the WCED. Its joint sponsorship by the federal and Manitoba governments and its international perspective has given it a different focus and set of challenges and opportunities compared the other three non-executive agencies discussed in this volume. The IISD has had the good fortune of strong leadership which has allowed it to turn what could have been crippling crises into creative opportunities. In the process, IISD has broadened its financial base and partnership networks. It has developed some world leading reporting tools for international negotiations and is very highly regarded in the constellation of international sustainable development organizations.

LAURA SMALLWOOD assesses the first dozen years experience of one of Canada's other institutional innovations, the Commissioner of Environment and Sustainable Development (CESD) and finds a mixed story. There is no question that the CESD's annual reports to Parliament outlining and critiquing various aspects of the federal government's sustainable development implementation performance have been very highly regarded. However, as Smallwood argues the structural constraint of placing the commissioner function within an audit office has fundamentally limited the scope of work that the various CESD's have been able to undertake. The chapter provides a detailed assessment of all the major eras, highlighting both the barriers and challenges and the achievements and opportunities. Both the domestic and the international domains of the CESD's work are evaluated. Based on this assessment and given recent legislative developments such as the passing of the Federal Sustainable Development Act, Smallwood proposes substantial institutional changes that would allow the CESD to perform to its potential.

Technological innovation will play a significant role in the "next industrial revolution" which is a central component of the move towards sustainable development. ANIQUE MONTAMBAULT evaluates the role played by Sustainable Development Technology Canada (SDTC) as a 21st Century instrument of public policy which invests public funds to leverage private sector investment to help build a sustainable economy in Canada. Montambault reveals the tortured political process involved in creating this arms length agency under the Liberals and shows its continued favoured status under the Conservatives. Under the guidance of key leaders, in its short life SDTC has become a substantial knowledge broker for sustainable development – linking back into federal departments and out to academia, other governments and especially industry. Working at arms length from government, SDTC invests public funds in promising clean technologies focused on greenhouse gas emission reductions, and clean, air, water, and soil. While SDTC presents an interesting Canadian success story, nothing is certain in the area of technology products and markets. Both markets and public policy provide barriers as well as opportunities for technological innovation, and Montambault explores critically the experiences to date of SDTC and future challenges on the road ahead.

NOTES

1 World Commission on Environment and Development (WCED), *Our Common Future* (Oxford University Press, 1987).

2 Ibid., 43.

3 Ibid.

4 W. Lafferty and J. Meadowcroft, eds. *Implementing Sustainable Development: Strategies and Initiatives in High Consumption Societies* (Oxford University Press, 2000).

5 W. Lafferty, "The politics of sustainable development: global norms for national implementation," *Environmental Politics* 5 (1996), 185–208; J. Meadowcroft, "Sustainable development: A new(ish) idea for a new century?" *Political Studies* 48 (2000), 370–87.

6 Consider, for example: The Millennium Ecosystem Assessment MEA, *Ecosystems and Human Well-Being: Synthesis Report* (Earthscan, 2006); and Intergovernmental Panel on Climate Change, *Synthesis Report of the Fourth Assessment Report* (IPCC, 2007).

7 Intergovernmental Panel on Climate Change, *Synthesis Report of the Fourth Assessment Report* (IPCC, 2007).

8 WCED, *Our Common Future*, 44.

9 *OECD Environmental Strategy for the First Decade of the 21st Century* (oecd, 2001).

10 Lafferty and Meadowcroft, *Implementing Sustainable Development.*

11 M. Janicke and H. Weidner, eds. *National Environmental Policies: a Comparative Study of Capacity Building* (Springer, 1997); K. Hanf and A. Jansen, eds. *Governance and Environment in Western Europe: Politics, Policy and Administration* (H Longman, 1998); J. Tatenhove, B. Arts, and P. Leroy, eds. *Political Modernisation and the Environment: the Renewal of Environmental Policy Arrangements* (Kluwer Academic, 2001).

12 M. Anderson and D. Liefferink, *European Environmental Policy: The Pioneers* (Manchester University Press, 1997).

13 A. Jordan, *The Europeanization of British Environmental Policy* (Palgrave, 2002).

14 N. Vig and M. Faure, *Green Giants? Environmental Policies of the United Sates and the European Union* (mit Press, 2004); R. Wurzel, *Environmental Policy Making in Britain, Germany and the European Union* (Manchester University Press, 2002).

15 J. Meadowcroft, "National sustainable development strategies: a contribution to reflexive governance?" *European Environment* 17 (2007), 152–163.

16 R. Steurer and A. Martinuzzi, "Towards a new pattern of strategy formation in the public sector: first experiences with national strategies for sustainable development in Europe," *Environment and Planning C: Government and Policy* 23 (2005), 455–72.

17 See K. Eckerberg, "Sweden: progression despite recession," in W. Lafferty and J. Meadowcroft, eds. *Implementing Sustainable Development: Strategies and Initiatives in High Consumption Societies* (Oxford University Press, 2000).

18 For the Environmental Objectives Council see: www.miljomal.nu/english/english.php.

19 L. Lundqvist, *Sweden and Ecological Governance: Straddling the Fence* (Manchester University Press, 2004).

20 United Kingdom, *Climate Change Act* 2008 (Department for Environment, Food and Rural Affairs, 2008). Available at: www.defra.gov.uk/environment/climatechange/uk/legislation/.

21 United Nations Development Program (undp) statistical update on December 18, 2008. Available at: http://hdr.undp.org/en/statistics/.

22 Yale Center for Environmental Law and Policy, and the Center for International Earth Science Information Network Columbia University, *2008 Environmental Performance Index: Summary for Policy makers.* Available at: http://sedac.ciesin.columbia.edu/es/epi/.

23 D. Boyd, *Unnatural Law* (University of British Columbia Press, 2003).

24 wced, *Our Common Future*, 57.

25 For a discussion of this point see: N. Höhne, C. Michelsen, S. Moltmann, H. Ott, W. Sterk, S. Thomas, and R. Watanabe, "Proposals for contributions of emerging economies to the climate regime under the unfccc post 2012," *Climate Change* 15.08 (German Federal Environment Agency, 2008).

26 United Nations Conference on Environment and Development, *Agenda 21* (United Nations Organization, 1992).

2 Why Aren't We There Yet? Twenty Years of Sustainable Development: A Personal View

DAVID RUNNALLS

I remember a conversation with a journalist a few months ago in which he breathlessly told me that environment was the top-of-mind issue in Canadian polls for the first time ever, supplanting national security, unemployment, health care and the like. And he wanted to know what I thought of that.

And that got me thinking that I had heard all this before. It was in 1988–89. And the issue continued to score in the polls until 1992.

At that time, Canada was the most advanced country on earth in terms of sustainable development. The Brundtland Commission had held hearings across the country, which drew huge crowds. The pioneering National Task Force on Environment and Economy had been established in the wake of Brundtland. It produced a report signed by Ministers, CEOs and civil society leaders with recommendations on how to integrate the environment and economics in decision-making, the most important insight of the Brundtland Commission, more formally known as the World Commission on Environment and Development (WCED).

This had been foreshadowed by Canada's official submission to the Commission during one of its many public hearings in Canada. "Unless the environmental sciences are routinely harnessed by economic scientists and decision-makers, the future of Canada – both economically and environmentally – is seriously threatened. Conventional economic analysis is the underpinning of all of the world's development decision-making. The inability

Text of a speech delivered to the conference "Facing Forward Looking Back" in October 2007.

of economics to take into full account the 'real' value of social and environmental assets has created enormous gaps in the ways societies define and reflect in decisions their well-being and the value of their future."

It then went on to admit that in Canada, "there is, however, almost no integration of economics and the environment at any level of government."

The Toronto Conference on the Changing Atmosphere had brought together more than 300 experts and policy-makers under the auspices of Gro Harlem Brundtland and Prime Minister Brian Mulroney. The rather alarming Conference Statement noted that "humanity is conducting an unintended, uncontrolled, globally pervasive experiment whose ultimate consequences could be second only to a global nuclear war." This was in 1988. This was not authored by Greenpeace or the Sierra Club. It was a group of climate scientists and policy people. So much for the theory that there is a great debate amongst the climate science community about whether the sources of increased CO_2 are largely anthropogenic.

The preparations for the Earth Summit in 1992, which was cleverly called for by the Commission to make sure that its report did not die on the vine, were well underway and negotiations had begun for what became the global conventions on climate change and biodiversity.

Canada's environmental non-governmental organization (ENGO) leaders had jointly agreed (a very rare phenomenon) on a statement, the Greenprint, and the country's industrial leaders had also come together. In a statement by the Business Council on National Issues, Tom d'Aquino said, "Reversing the deterioration of the environment on a global basis is the most pressing challenge facing Canadians and the world."

The momentum continued with the establishment of multi-stakeholder Round Tables in each province and territory, along with the National Round Table on the Environment and the Economy. The high water mark of the Round Tables came with a meeting in Winnipeg of all the Round Tables convened by Manitoba Premier Gary Filmon. CEOs, Ministers and ENGO leaders spent two days comparing notes and urging action.

The creation of the International Institute for Sustainable Development (IISD) was announced by the federal Environment Minister Lucien Bouchard at the first meeting of Globe, in Vancouver. We and Globe both celebrated our 15th anniversaries in 2005.

Finally, we had the Green Plan, produced over a number of years and after a seemingly endless series of consultations. Although watered down considerably because of interdepartmental suspicions that Environment Canada was seeking to transform itself into a rival to some of the central agencies, and because of the political shenanigans of its original proponent, Lucien Bouchard, the plan was still a world-leading document. Its chapter on decision-making, relegated to the end of the report by a PR expert who felt that the public did not want to hear about a reorganizing of government thinking as the lead chapter, remains a benchmark on the subject.

So we had everything in place to move forward with sustainable development, which is the hallmark of the Brundtland Report. Needless to say we have been going backward, or perhaps more generously, sideways, ever since.

I guess my story begins in 1972 with the Stockholm Conference (the United Nations Conference on the Human Environment) and the publication of *Only One Earth*. The UN had planned the world's first global conference on the environment. By 1970, it was clear that the developing countries had little interest in the idea. In fact, many of them viewed environmental protection as a threat to their own plans for development. In an effort to rescue the conference, the Secretary General of the UN sought a Secretary General for the Conference who commanded the confidence of developing countries. He chose Canada's Maurice Strong, then the President of the newly formed Canadian International Development Agency (CIDA). Maurice undertook a number of confidence-building initiatives with the Third World. Chief among them was his decision to ask my boss, the British writer and economist, Barbara Ward, to write the theme book for the conference. I was the uninformed research assistant who organized much of the research for the book. Only One Earth, which Barbara co-authored with the Franco-American biologist René Dubos, was an instant hit. I reread it in 2002 on the 30th anniversary of Stockholm and it still reads pretty well.

Stockholm was a landmark event – more journalists attended than were present at the Munich Olympics later that year. Maurice got developing countries to come and to participate in a major way. Indira Gandhi, the Indian prime minister, delivered the speech of the Conference. She made the link between a deteriorating environment and the growth of poverty in the starkest terms.

Stockholm led to the establishment of many environment ministries in OECD countries, greater public expenditures on the environment and a good deal of legislation.

Despite this progress, it became clear that the environment was still not a major priority for most countries. And the Global 2000 report to the president, produced in the Carter White House and strangled in its cradle by the Reaganites, made it clear that the earth's natural systems were in very serious trouble indeed.

Accordingly, Canada pressed hard in the Governing Council of the UN Environment Programme for the creation of a global commission to examine the relationship between environment and development. Born out of a Canadian initiative, the World Commission on Environment and Development had two Canadian members: Maurice Strong, who appears with great regularity in any international environmental narrative, and Jim MacNeill, the secretary general and the hand that held the pen, as well as the guiding force behind the report.

I need to put the report in its international political context. The Commission came along at a time when both East-West and North-South relations were poor and previous World Commissions, such as that on development headed by the former West German chancellor, Willy Brandt, had conspicuously failed to reach unanimous agreements among their members. And the WCED was geographically representative, as they like to say in the UN. It contained a Soviet scientist, an unreconstructed Chinese Communist and a Reagan Republican, among others. So the omens were not good.

But they did produce a remarkable report. When I agreed to give this talk, I got *Our Common Future* (the Brundtland Report's official name) off the shelf and reread it. While I was not embarrassed when I reread the book I had helped Barbara Ward to write in 1972, I was bowled over by the Brundtland Report. Change a few dates and a few references and it is as timely today as it was in 1987.

Last fall, Jim MacNeill gave a talk at the University of Ottawa in which he spoke of the Brundtland Report. And since he wrote it, I will let him summarize it:

Twenty years ago, in Chapter 1, we described a world threatened by interlocking crises. We spoke of rising levels of population and a spiraling growth of megacities in the Third World along with massive projected increases in consumption in the First; of increasing levels of poverty and inequity within and between nations; of continuing huge transfers of wealth from the poor to the rich built into grossly inequitable trading relations; of unsustainable increases in the consumption of our natural capital, our soils, waters, and forests; of the destruction of species; and of the growing menace of climate change. We pointed out that these environmental syndromes presented a *"threat to national security and even survival"* greater than any military threats then on the horizon.

If that doesn't sound familiar, look at yesterday's headlines – or even today's – or at today's best-seller lists?

We pointed out that *"most efforts to maintain human progress [to] meet human needs and [to] realize human ambitions are simply unsustainable in both rich and poor nations"* and if we continued on these paths we would *"threaten ecological collapse."*

One final quote that I find equally germane 20 years later. "We act as we do," we said, "because we can get away with it: future generations do not vote; they have no political or financial power; they cannot challenge our decisions … We are … rapidly closing the options for future generations."

And so, after much debate and analysis, we concluded that we must change course. *"Humanity,"* we said, *"has the ability to make development sustainable,"* but we don't have much time – some decades at the most. And so we called for an urgent and rapid change worldwide to more sustainable forms of development.

Now this was pretty powerful stuff and it attracted an enormous amount of media attention, not least because the report was not only tough, but all of the Commissioners, including the Soviet member and the unreconstructed Chinese Communist, as well as President Reagan's former EPA director, signed it.

Maurice Strong used the oxygen created by the report to energize the preparations for the 1992 Conference in Rio. Since the Stockholm Conference in 1972, the UN had staged two-week conferences on all kinds of subjects. Many of these were inconclusive or indecisive and the format was tired. Maurice decided to up the ante. He egged on those negotiating the climate change and biodiversity conventions to speed up their timetables so that the conventions would be ready for signature by the time of the Conference. And then he upped it again by turning it into a summit, rather than a meeting of Environment Ministers. This reflected one of the principal learnings of the Brundtland Commission – sustainable development had to be a top-down process in governments and corporations. Jim's idea that the national budget would become the government's most important annual statement of sustainable development meant that the Head of Government had to take this concept seriously or it would not go to the heart of the decision-making process. Just as Ed Woolard of Dupont once described the term CEO as Chief Environmental Officer, now the ministers of Finance would become ministers of Finance and Sustainable Development. And this would be legitimized in Rio by a massive Round Table lunch of some 120 heads of government who would be asked to sign a ringing declaration, as well as the two conventions.

And Canada played an important role at the Conference. Maurice was the star of the show. Brian Mulroney saved the Biodiversity Convention from destruction at the hands of the semi-articulate US vice president, Dan Quayle (remember the potato man?), by agreeing to sign it. This encouraged the Europeans to stand up to American pressure. After all, if the great Irish tenor who joined Ronald Reagan in singing "When Irish Eyes Are Smiling" could stand up to Bush 41, why so could they.

And the delegation was led by the then Environment Minister Jean Charest, who had the time of his life. And then the Conference came to an end. And they all came home. And very little happened.

In 2001, I was asked to chair a multi-stakeholder panel to prepare the Canadian National Report for the World Summit on Sustainable Development (the Johannesburg Summit of 2002). I won't go into the gory story of how the report was suppressed by the PMO after it had been written. Our report found that there had been some progress. The office of the Commissioner for Environment and Sustainable Development had been created and was functioning well. The National Round Table on the Environment and the Economy and my own institute, the International Institute for Sustainable Development, had been created. An ill-prepared delegation had gone to Kyoto and negotiated the protocol. Canada had taken the lead on the POPs convention. And many companies in Canada had begun to shift their attention to the development of corporate environmental policies, which went beyond mere compliance with environmental rules and regulations, and were undertaking proactive environmental reporting initiatives to their stakeholders. But compared with the scale of the challenges identified by Brundtland, not nearly enough had happened and Canada ranked near the bottom on OECD surveys of the effectiveness of national environmental policies.

Sustainable development, once one of Canada's strong points, had taken root in Europe and several countries had made substantial progress on the integration of environment and economy and the creation of high-level bodies to begin the incorporation of sustainable development into mainstream decision-making. Perhaps most notably, sustainable development had been embedded into the Treaty of Amsterdam, the EU's basic constitution at the time, as one of the guiding principles of the Union. It had even become one of the central principles of the Marrakech Declaration, the final document of the Uruguay Round, which created the World Trade Organization.

WHAT HAPPENED TO US?

I guess we all have theories about this one. Mine goes something like this:

1 Constitutional crises: Remember the ill-fated Charlottetown accord? The politicians who were part of the Canadian delegation at the Earth Summit in Rio returned to Ottawa in June and were immediately caught up in the negotiations for the accord and the debate over the referendum, which failed in the fall of that year.
2 Eco-fatigue: In some ways Rio was a victim of its own success. It had attracted a massive press corps; there had been whole issues of *The Economist*, *Time* and *Newsweek* devoted to the event. Because the heads of government were there for the second week, all the superstar political reporters tagged along, and their coverage augmented that of the environmental journalists who had already been filing for a week.

   Environmental historians have written a good deal about the waves of public support for environment and have traced these waves from the creation of the U.S. and Canadian national parks systems. And Rio was likely the end of such a wave, at least in North America.

   And it soon became apparent that Rio had not really delivered as much as all the hype seemed to indicate. Much was written, by Jim MacNeill and me, as well as others, about the Rio bargain. Because of the parlous state of North-South relations, Rio never really faced up to the challenge of sustainable development. Instead of a sustainable development agenda, the conference really faced two agendas: (1) the Canadian and other OECD countries' agenda of climate change, loss of biodiversity, forest loss and marine and fisheries issues; and (2) the developing country agenda of trade liberalization, increased aid, access to advanced technology and the famous principle of common but differentiated responsibility. The cracks were papered over by an implicit deal that the developing countries would pay attention to our agenda if we increased official development assistance considerably. Instead, most developed countries actually decreased their aid flows, including Canada after CIDA was a serious victim of Program Review. The failure of that bargain still plays a role in the reluctance of many developing countries to assume targets under the Kyoto process. They simply do not trust us.

So Rio did not have any obvious deliverables for the public, except for Kyoto, which flew below the public radar screen in Canada until the Johannesburg Summit in 2002.

3 Recession and the war against the deficit: And this should not be underestimated. The Mexican peso crisis brought the state of Canada's shaky public finances into stark relief. The Chrétien government was determined to get the deficit under control. Despite all of the good things said in their initial "Red Book" about sustainable development, the Environment Canada and CIDA budgets were cut dramatically and these cuts escalated when the Green Plan funding expired as well. Since the Green Plan funding was time bound, Finance did not have to count it as a cut. So, while the Environment Department's cut was proportional to those of some other departments, it was actually as high as 30 per cent or more, according to some estimates.

4 Financial stringency also affected the corporate sector: And the first wave of corporate believers who had been instrumental in the success of the Round Tables had moved on. They were replaced by more traditional bottom-line-oriented CEOs, many of whom were unaware of the sustainability debate and viewed environment as a cost, which could be ignored until times were better. This was one of the factors in the death of the Round Tables, one of Canada's genuinely innovative responses to the challenges of environment/economy integration and unique fora for bringing ENGOs, politicians and business leaders together. Corporate retrenchment left little time for luxuries such as the environment.

5 Major cuts occurred at the provincial level as well: In many cases, these were much more serious than at the federal level. For example, the Ontario Ministry of the Environment has still not recovered from the cuts of the Harris era.

6 And then we have that all-time favourite, political will: Or rather, lack of it. With some notable exceptions such as David Anderson and Charles Caccia, few were prepared to speak up for the environment.

Our world has changed dramatically over the past two or three years. According to Globescan, another Canadian invention, Canadians are more concerned about climate change than the citizens of any other developed country. And it is not inconceivable that we could still have a defeat of the government over climate policy. This is part of a worldwide trend, which is perhaps most obvious in Europe and which culminated in the award of the Nobel Peace Price to Al Gore and the Intergovernmental Panel on Climate Change (IPCC) last year. Political opinion is even changing in the U.S., where all three remaining major Presidential candidates have strong positions on climate change.

What do we have to do in order to restore Canadian respectability, if not leadership, on sustainable development?

1 We need to remember the number one issue of Brundtland: And that is that the world's environment and its economy are so closely linked that policies on one that ignore the other are bound for failure. We need to integrate environment into economic decision making at all levels of government and in the private sector. This integration can only be done with strong leadership from the top. We need ministers and bureaucrats to follow the existing Cabinet directive requiring strategic environmental assessments of all major policy decisions before they are taken.

2 We need at least a federal sustainable development strategy: I would like to suggest a national strategy, developed through public consultation, both through electronic and other means. But if not possible, at least a government-wide strategy. The lack of such a strategy has hamstrung the work of the commissioner for Environment and Sustainable Development from the beginning. A collection of departmental reports does not an overall strategy make.

3 We need to develop an economic policy that promotes sustainable development: cities, provinces and Ottawa itself need to keep environment top-of-mind as they develop budgets. Stand-alone environmental policies and projects can be useful, but until the environment is truly integrated throughout all economic policy-making, real change will be stunted.

4 We need a national conversation about energy policy: As the prime minister has said repeatedly, we are an energy superpower. We need to act like one.

5 We need a climate policy, which is acceptable to Canadians as a whole: It looks as if we are moving toward a crazy quilt of federal and provincial policies at a time when I suspect that the United States will be moving in the opposite direction.

6 We need to do something about adaptation to climate change: As any Northerner can tell you, climate change has already begun in a major way north of 60. Adaptation is a complex subject, ranging from the need for new patrol vessels to reinforce Canadian sovereignty in the Arctic, to flood defences for Surrey and Richmond, BC.

7 We need a more sustainable approach to natural resource management: Canada's "natural capital" is one of its greatest assets. Although our management of fisheries, soils, water and forests seems to be slowly improving, we have a long way to go.

8 Developing countries need to be part of the solution: Climate change is a truly global problem. A tonne of CO_2 emitted in Vancouver has the same

impact as a tonne emitted in Beijing or Brasilia. Canada needs to take a leading role in helping developing countries to develop their own sustainable energy paths.

9 Reform of international environmental governance: We need to have fora where real agreements can be reached. Although the 190 some odd countries present in the December 2007 climate talks in Bali all have a right to be part of the solution, 15 countries are responsible for the vast majority of the world's emissions. We need to find a way to get those 15 to agree on policies to curb emissions, to price carbon and to create new technologies, before the other 175 are brought into the debate.

10 Finally, it goes without saying that we have to engage and energize young people: Climate change is the biggest challenge to sustainable development. And if we do not take strong actions within the next 10–15 years, the world will become an awful place to live. I may not see it within my lifetime, but my children will – and they cannot sit idly by and passively watch it happen.

3 Institutionalizing Sustainable Development: The Role of Governmental Institutions

GLEN TONER AND
FRANÇOIS BREGHA

In an ideal world, scientific investigation and analysis inform political debate and lead to policy changes. The 1980 World Conservation Strategy (wcs) was a good example. It employed the leading scientific research to take the pulse of the global ecological system and concluded that human kind was having a significant negative impact on the environment. Poverty, population pressures, social inequity, and the terms of trade were resulting in significant habitat destruction. The diplomats at the United Nations heeded the warning of the scientists as they debated the wcs. The result was the establishment in 1983 of the World Commission on Environment and Development (wced). The wced's landmark 1987 report *Our Common Future* (ocf) (also known as the Brundtland Report) weaved together the social, economic, environmental, and cultural issues raised by economic development and environmental degradation and proposed global solutions. It popularized the concept of sustainable development (sd) that had been introduced by the scientists in the wcs. In one of its most insightful passages, ocf argued "sustainable development is not a fixed state of harmony, but rather a process of change in which the exploitation of resources, the direction of investments, the orientation of technological development and institutional change are made consistent with future as well as present needs."[1] In its Call for Action, the wced implored governments to reorient their policies, programs, and institutions to ensure that sd became an integral part of the mandate of all agencies of government.[2]

There is, in fact, a direct line of causality between the recommendations of ocf and transformations to the policies and institutions of the Government of Canada over this period. Public policy plays an important role in the integration of sᴅ into the practices of firms, citizens, and of governments themselves. Through their wide range of fiscal, regulatory, and program activities, governments monitor changes and threats to the environment, raise public awareness, set societal goals, create patterns of incentives for firms and individuals, and invest in analytical and management tools to reduce adverse impacts, remediate problems, and prevent future ones. Indeed, a key recommendation of oᴄꜰ was to shift the orientation of government programs from an after the fact "react and cure" approach to a proactive "anticipate and prevent" approach. While our focus is on the Canadian federal government, these functions apply to governments at all levels and, given the inevitably cross boundary character of most sᴅ issues, international institutions and processes as well. We have both studied this transformative change process closely and published detailed assessments of the achievements and failures of the period; the first part of the chapter provides a high level review of the historical engagement in order to propose changes to improve performance over the next 20 years in the second half.

LOOKING BACK: 1987–2007

Framework Policies and New institutions

A review of the Canadian government's performance in institutionalizing sustainable development over the last 20 years needs to be placed in its historical context. The Brian Mulroney–led Conservative Government was first elected in 1984 and paid scant attention to environmental issues during its first mandate. The first minister of the Environment was weak and did not last long in office. The government toyed with the idea of abolishing the department of the environment and in the end made serious budgetary cuts which hurt the wildlife service in particular. While Environment Canada had been around for a decade and a half, it had little regulatory authority, very limited policy capacity, and was universally regarded as a junior ministry. Yet, by the end of the first mandate a strong minister was in place and the Conservatives were moving environmental issues into the mainstream.[3] When *Our Common Future* was published in 1987 and introduced the idea of sustainable development, Canada was experiencing a period of rising public interest in environmental issues. Indeed, the environment as an issue topped public opinion polls in 1989–90 as it did again in 2006–07. This was driven by a combination of catastrophic industrial accidents, breakthrough scientific discoveries, increasingly active international policy debates and treaties, a growing and robust environmental ɴɢᴏ movement, and for the first time an attentive print and broadcast media. In his second term, Mulroney appointed a senior minister to the portfolio, created a cabinet committee on the environment and strengthened spending programs, scientific capacity, and international leadership in this area. There was even talk of making Environment Canada a central agency to oversee the government's sd approach. While the traditional bureaucratic power structure resisted this and other changes, Mulroney continues to take great pride in his innovative and progressive response to *Our Common Future* and he has, in fact, been acknowledged as Canada's greenest prime minister.[4]

The wced visited Canada in May 1986 and met with many groups of Canadians. Among them was the ministerial body, the Canadian Council of Resource and Environmental Ministers (ccrem). After this meeting, the ccrem created a National Task Force on Environment and Economy. This multisectoral body was comprised of ministers, industrial, university, and environmental group leaders. The members were so positive about the innovative dialogue it created amongst traditional adversaries that its most important recommendation was to encourage Mulroney to institutionalize its process for reaching integrated solutions to environment-economy issues by creating a National Round Table on the Environment and Economy.[5] It also recommended that similar advisory bodies be set up by provincial premiers and territorial leaders. A series of additional policy and institutional innovations emerged in the decade after 1987 which led Canada to be generally regarded at the time as a global leader in sd governance.[6] The initial policy framework response was the Mulroney government's 1990 Green Plan, the first comprehensive federal environmental policy ever. From the 18th Century to the 1960s Canada was developed under a 'Frontier Economics' paradigm. This was followed from the 1960s to the late 1980s by the resource management/environmental protection paradigm. The Green Plan self-consciously attempted to anchor its approach within the new sd paradigm. "Sustainable development is *what* we want to achieve. The Green Plan sets out *how* we are going to achieve it."[7] The Green Plan was led by Environment Canada but resisted strongly by the finance and industry departments, and never gained a strong ally in the central agencies. While it never achieved the aspirations of its political and bureaucratic architects, it did lead the government's agenda for a couple of years, pumped an unprecedented amount of new money into environmental programs for a time, and introduced and supported several new or evolving tools.[8] The highlight of this period was the 1992 Rio Earth Summit where Mulroney lauded the Green Plan as an example of sd planning and Environment Minister Jean Charest led daily "Team Canada" briefings for all the Canadian delegates, governmental and non-governmental. In fact, Canada did receive considerable international praise for the introduction of the Green Plan with one analyst even calling it the "Mother" of green planning.[9]

In opposition, the Liberals were highly critical of the Conservative's Green Plan, charging it with not going far enough in changing the decision-making system to institutionalize an SD framework.[10] They dedicated a chapter of their 1993 election manifesto to SD and proposed several additional program and institutional modifications which they later introduced. The first Liberal budget after the election created a task force to undertake a review of federal taxes, grants, and subsidies, in order to identify barriers and disincentives to sound environmental practices. The Task Force report identified a number of both short and long term measures that would have enhanced the use of market based instruments. The Liberals created the first Standing Committee on Environment and Sustainable Development (SCESD) in the history of Parliament.[11] Industry Canada and Natural Resources Canada both integrated commitments to SD into their statutory mandates.

The 1995 Guide to Green Government (GGG), which was signed by Prime Minister Chrétien and the entire cabinet, outlined the policy context for the creation of a Commissioner of Environment and Sustainable Development (CESD). The GGG stated that "achieving sustainable development requires an approach to public policy that is comprehensive, integrated, open and accountable. It should also embody a commitment to continuous improvement."[12] To institutionalize this approach, the government established a legal requirement that departments create and implement sustainable development strategies (SDS) every three years and charged the CESD with monitoring departmental performance. The CESD reports annually on various dimensions of the government's performance.[13] In 2001 the government created Sustainable Development Technology Canada (SDTC) to expedite the commercialization of innovative environmental technologies to reduce greenhouse emissions and improve air quality and to strengthen the performance of Canadian industry.

Institutions

Interestingly, the four new institutions (CESD, SDTC, NRTEE, International Institute for Sustainable Development) created as parts of the government's response to SD are all outside the executive branch. Indeed, all four are arm's length from government in terms of operations yet, in part as least, dependent on government for funding. The four new organizations have all added real capacity to the national SD intellectual infrastructure. Yet, the marginalization of SD institutions outside the executive branch is also part of the problem. To be meaningful and to alter the decisions coming out of the government, SD components have to be built into the mainstream units and processes of key line departments. As the CESD performance audits of the past decade have shown, this was done in some cases, with SD programming integrated into departmental business planning. Treasury Board assessment criteria and systems related to the parliamentary reporting system were created. Public Works and Government Services, in concert with other key departments, tried at various times to exercise a government wide authority over green procurement. But to be brutally honest, the government never really figured out how to make this work. Far too often, the units within the department charged with integrating the work of various groups were too junior or too far removed from the epicentre of departmental decision making. While there were of course exceptions throughout the large government apparatus, most commentators agree that not enough senior managers became engaged. Without the systematic integration of sD principles and values within bureaucratic processes, it is not surprising that departments would make robust and creative commitments within their sDs and fail in implementing them. This disconnect between what they know has to be done and what they do has resulted in serious underperformance by the mainstream federal departments.[14]

The cesD is the most closely connected to government operations through its ongoing performance audits of government programs and its annual review of the implementation of departmental sDs. The creation and ongoing operation of the cesD in the Office of the Auditor General (oAG) is the most contentious organizational issue amongst the four. The scesD initially recommended that the cesD be established as an independent Officer of Parliament.[15] In large part because of vociferous opposition from within the bureaucracy the government opted to place the cesD at a second level within the oAG. The work of the cesD has been highly regarded by the policy community since 1997 but the built-in limitations of its mandate – it is an auditor, not a champion for new policies or processes – has continued to haunt its operations and was recently the subject of a Parliamentary Review before the scesD.[16] (See chapter 11 of this volume)

As the newest of the group, sDTC has experienced a different set of growing pains. Its mission is to act as a primary catalyst in building sD technology infrastructure in Canada. Successive budgets have continued to add new resources, including a substantial infusion of $500 million in 2007. Because it assesses and invests in private sector projects it has a stake in protecting the privacy of its clients' intellectual property. Changes introduced by the Harper Government's *Federal Accountability Act* broadening the application of access to information rules and extending performance audits of foundations by the AG have raised potentially significant viability issues for an ostensibly arms length organization like sDTC. (See chapter 12 of this volume)

As mentioned, when the NRTEE was set up in 1989 the goal was to bring together leaders with varied experience to explore diverse options and integrative solutions to sustainable development challenges. Initially federal ministers participated as members. However, by the time the NRTEE received its own legislative base in 1993, ministers were no longer participating and Prime Minister Chrétien largely ignored it. Also by this time the focus of the NRTEE's work had shifted from seeking consensus based solutions at the front end of policy debates to undertaking research projects to establish the "State of the Debate" in key area areas of sustainable development. The NRTEE has always struggled with the tension between wanting to be an independent reflection of leading thinking or an inside advisor whose ideas were used by politicians and the public service. As an organization the NRTEE has gone through several shifts in its orientation and each new government has also had to determine how to use it and its advice. While the Harper government has demoted the reporting relationship of the NRTEE from the prime minister to the minister of the Environment, it has at the same time asked the NRTEE to explore key long term climate change related research which may have a real influence on government policy.[17] (See Chapter 9 of this volume)

Green Plan funding contributed to the creation of the International Institute for Sustainable Development (IISD) in partnership with the government of Manitoba. The IISD has the most distance from the federal government both in terms of mandate and location. Its headquarters are in Winnipeg, though its staff of 150 is located in 30 countries. Its broad mandate is to promote change towards SD through policy research and communications with a broad range of decision makers both within Canada and abroad. Its work spans the full range of SD concerns from local ecosystem issues on the Canadian prairies to the broadest international issues. As a non-profit body, it receives core operating support from the governments of Canada and Manitoba as well as program funding from a broad range of domestic and international bodies, including other national governments. IISD has created a particularly interesting niche in reporting on international negotiation processes. It seeks through information-sharing functions to build both its own institutional capacity and partner organizations in the developing world. IISD's research and reporting program encompasses the trade and investment, measurement and assessment, climate change and energy and sustainable natural resource management areas. (See chapter 10 of this volume)

Policies and Instruments

Besides creating four new institutions, the federal government also deployed a number of new policies and instruments to promote sustainable development. These can be broadly categorized according to their focus. Most, such as regulations, voluntary approaches and economic instruments are intended to change the behaviour of Canadians in ways that protect the environment. Others, such as strategic environmental assessment and leadership by example, are primarily intended to improve internal government decision-making. Finally, some instruments, such as sustainable development strategies and information serve a dual purpose in both changing public behaviour and improving policy-making. Let us examine the federal government's record in deploying these instruments.

Instruments with external focus

Regulations   Over the last two decades, federal environment and sd policy has developed against a shifting background of changing public opinion, large government budget cuts (in the mid-1990s) offset in part by significant new resources in the last decade, evolving science (the developing consensus around climate change) and the ebb and flow of federal-provincial politics. Governments around the world favoured a regulatory approach to environmental protection when the environment first emerged as an important policy field in the 1970s. The Canadian government was no exception. In the last 20 years, it has considerably strengthened its regulatory framework for environmental protection. While the government put in place several noteworthy environmental regulatory initiatives over the past two decades, three in particular stand out for their broad scope:

- The *Canadian Environmental Protection Act*: introduced in 1988 and extensively revised in 1999, cepa provides comprehensive regulatory authority to protect the environment and human health by, among other things, controlling toxic chemicals, waste disposal at sea, the products of biotechnology, hazardous wastes and the pollution from transportation sources.
- The *Canadian Environmental Assessment Act* came into force in 1995. The Act sets out different types of environmental assessment for projects over which the federal government has authority, depending on their complexity and anticipated impacts.
- The *Species at Risk Act*: passed by Parliament in 2002 after many years of rancorous debate, sara's objective is to prevent wildlife species from becoming extinct, and to secure the necessary actions for their recovery. While federal intervention has spurred several provinces to greater efforts to protect their wildlife, the federal government's ability to act decisively is ultimately limited by provincial jurisdiction over land and resources, one reason why the government is not meeting the legislated deadlines for the implementation of recovery strategies for threatened species. The Act therefore emphasizes collaboration with other governments, Aboriginal Peoples and land-owners to achieve its objectives.

The 2006 election of Stephen Harper signalled another shift in federal environmental policy. While Canada's new government abandoned the climate change targets embodied in the Kyoto Protocol, it announced its intent to make a greater use of regulatory approaches to control both air and chemical pollution. Its proposed regulatory framework for air emissions is arguably one of the most significant regulatory initiatives the federal government has launched in any field. While this framework proposes ambitious air quality objectives designed to place Canada among world leaders, the long time that has elapsed since its announcement without the tabling of detailed regulations suggests that this initiative is likely to be watered down. The Harper government in the last half of 2007 said it would not abandon the UN based international process to establish a post-Kyoto period regime but it also announced that it was joining the Asia Pacific Partnership made up of six countries, most of whom have not taken on emission reduction targets under the Kyoto Protocol.[18]

Under the Conservative government, Canada has taken the view that any global climate change treaty needs to include binding emissions reduction commitments from all major emitters, even if these are developing countries that have historically not contributed significantly to the increase in CO2 concentrations in the atmosphere. The government has also espoused medium and long-term reduction goals for greenhouse gases of 20% by 2020 and 50% by 2050. In the absence of detailed implementation plans and short term targets, these goals must be seen as aspirational rather than firm commitments.

At the same time as the federal government was putting this regulatory framework in place in the 1990s, the limitations of a purely regulatory approach were becoming more evident to policy-makers (viz., cost, limited effectiveness, limited incentives for innovation), prompting a growing interest in Canada as well as elsewhere in the deployment of complementary instruments to achieve environmental goals. Two such instruments in particular deserve mention here, one for its too-trusting acceptance and the other for its neglect. These are voluntary approaches and economic instruments.

Voluntary approaches   Voluntary approaches cover a wide gamut of environmental protection efforts ranging from challenge programs, environmental performance agreements between government and industry, to sector-wide or company-specific codes of conduct. Properly designed, such instruments can constitute a useful complement to regulatory approaches and, indeed, their interplay with government regulation implies that they are not always truly "voluntary."[19] Over the last two decades, the government has encouraged various industry sectors to improve their environmental performance voluntarily. It signed a series of memoranda of agreement and environmental performance agreements with some companies and industry associations, challenged companies to reduce their polluting emissions (the Accelerated Reduction and Elimination of Toxics program or ARET) and supported companies wishing to implement an environmental management system or report publicly their environmental actions and impacts.[20]

As a result of such initiatives and many other factors, including market pressures, government technological assistance, changing public opinion and the threat of government regulation, the overall environmental performance of Canadian industry has measurably improved over the last two decades. While voluntary initiatives remain an important element of the government's environmental "protection toolkit," these approaches have recently lost some of their lustre as their effectiveness, particularly as alternatives to regulation, has been increasingly questioned.[21]

Economic instruments   In 1992, the Government of Canada published a discussion paper on *Economic Instruments for Environmental Protection*. Since this publication, Canada has made only limited progress relative to many OECD countries towards incorporating such measures into its environmental conservation and protection toolbox.[22] The economic instruments the government has favoured have primarily been subsidies (e.g., to renewable forms of energy, for energy efficiency, for the donation of ecologically-significant lands) as opposed to pollution taxes or tradable permit schemes. In the 2007 budget, the government announced a system of fees and rebates to encourage consumers to purchase more efficient automobiles (since discontinued). Canada's slower progress in using economic instruments relative to other countries may mean it is not addressing its environmental problems as cost-effectively as it should. Closing this gap may be necessary to enable Canada to manage many of the complex environmental problems that are closely related to current systems of industrial production and consumption while continuing to stimulate innovation and competitiveness.[23]

Canada's limited use of economic instruments is the result of numerous factors. These have included (i) design challenges (e.g., the difficulty in predicting adjustment costs, concerns about distributional impacts, or uncertainty about the precise environmental impacts or the amount of revenue the measure will generate); (ii) strong public antipathy to "new taxes;" (iii) the challenge in designing "one size fits all" economic instruments in light of Canada's geographic, cultural and economic diversity; (iv) the difficulties inherent in diverging significantly from U.S. approaches to some environmental issues given the high economic integration between the two countries; and (v) persistent opposition by the Department of Finance to the use of taxes to pursue environmental objectives.[24]

In contrast to the federal government's highly cautious approach, British Columbia launched in early 2008 Canada's first "green tax shift" involving the imposition of escalating carbon taxes and the simultaneous reduction of other taxes. Later in 2008, the federal Liberal opposition adopted this idea and made it the main plank of its electoral platform. Both initiatives have 'forced' a public debate over the merits of pricing environmental "bads", such as carbon emissions and offsetting the increased government revenues by tax cuts in other areas, so that the overall effect is fiscally neutral. While the Conservative government attacked the Liberal proposal as a tax grab, its own proposed "cap and trade" regime for carbon emissions would also impose a price on carbon, at least for industry. While the quality of the Canadian debate nationally and in BC has been disappointing, 2008 may in time represent the crossing of a major threshold in legitimizing the use of economic instruments to achieve environmental ends.

Instruments with internal focus

Strategic Environmental Assessment (SEA)   SEA is a decision-making tool to integrate environmental considerations into government policy and programs. Its overall objective is to lead to the design of policies that promote more environmentally sustainable forms of development. In Canada, SEA is guided by the 1999 Cabinet Directive on the Environmental Assessment of Policy, Plan and Program proposals (as revised subsequently). A concerted effort has been made to build federal capacities to conduct SEAs over the last eight years through both departmental and government-wide training, the provision of various tools, the creation of networks of practitioners and the development of customised departmental guides. As a result, SEA is slowly taking root in several departments although the extent to which it influences decision-making is mixed.[25]

Leadership by Example   The government has invested considerable effort to greening its own operations over the last 15 years, starting with the 1992 Code of Environmental Stewardship. Having initially launched a variety of programs to address specific issues, it eventually consolidated these into a limited number of priorities under a newly-formed Office of Greening Government Operations in the Department of Public Works and Government Services. In 2002, as part of the Climate Change Plan for Canada, the Government announced that all its new facilities would exceed the Model National Energy Code and that all new federal housing units would meet R2000 guidelines. It also announced its intent to complete energy efficient retrofits in a further 20% of its buildings and to procure only goods and equipment that met the energy-efficient ENERGY STAR® standard. It made additional commitments to increase the proportion of low-emitting vehicles in the federal fleet, and for the government to be a first purchaser of next generation power technologies and energy sources. The government's efforts have been successful in many respects (e.g., the retrofit of 7500 buildings, generating $38 million in annual energy cost savings) although they have progressed slowly in some areas, such as green procurement where the CESD reports in 2005 and 2008 that the government was still not using the potential of green procurement as a tool to achieve sustainable development objectives.[26]

Instruments with dual purpose

Sustainable Development Strategies (SDS)   The government introduced the requirement that most federal departments would have to develop a sustainable development strategy and update it every three years in 1995. Departments have prepared four rounds of such strategies since then. These strategies have had a number of positive impacts: they have greatly raised awareness of SD issues within departments and across government; they have legitimised SD concepts and raised them to an accepted part of the policy discourse; departments have increased their capacities to analyse the environmental dimensions of their programs; the strategies have also led departments to think about how to promote SD and forced them to articulate action plans to address specific issues.

However, with some exceptions, the strategies remain poorly integrated with departmental business plans and focus heavily on "business-as-usual" programming. They are largely introspective in nature, focusing most of their attention on greening departmental operations, raising staff awareness and building internal capacity through background studies and research. They also show a predisposition to "soft" policy instruments (e.g., exhortation, voluntary initiatives) as opposed to "hard" instruments (e.g., regulations, taxes) and do not address most of the root causes contributing to the unsustainable lifestyle of Canadians, such as high resource consumption, land degradation, toxic pollution and loss of biodiversity. In 2005, while acknowledging some improvement, the CESD concluded that the majority of strategies focus more on process, activities, and outputs than on long-term results and most still did not include measurable targets with clear deliverables and deadlines.[27]

In June 2008, Parliament passed a private member's bill (C-474) to force the federal government to articulate government-wide sustainable development targets every three years that would become the basis for the preparation of departmental implementation strategies. The first set of targets is due in June 2010 and the departmental implementation strategies a year later. While the *Federal Sustainable Development Act* gives the government full discretion in defining these targets, it should result in a more focused federal effort than the past decentralized approach has yielded.

Environmental Information   In the early 1990s, Canada was widely recognized as a world leader in environmental information, as exemplified by its initiatives in environmental indicators, state of the environment reporting, and the examination of how to incorporate environmental and natural resource considerations in the national accounts. The government unfortunately cut this capability as part of its spending reductions in the mid-90s deficit fight. As a result, while Canada has continued to invest in environmental information (e.g., the development of six sustainable development indicators)[28] it has not developed a set of integrated indicators that would enable it to measure its environmental performance.

During this period, the government also invested in the development of the National Pollutants Release Inventory (NPRI), a publicly searchable database focusing primarily on industrial air emissions. Such a database provides both valuable information to decision-makers about the scope and origin of certain environmental problems and encourages individual companies to reduce their emissions to avoid unwelcome publicity. While the value of such inventories as a tool of public policy has been clearly established, the NPRI has not fulfilled its full potential because of data quality issues, ongoing definitional changes and insufficient attention to client needs.[29]

An Assessment of 1987–2007

In broad terms, the Canadian government can boast several important achievements over the last twenty years. As shown above, new institutional capacity has been developed, policy frameworks have been created and policies and instruments developed and deployed. At the program level we have seen the categorization of all chemicals in use in Canada – a world first – the creation of several new national parks, improvements in air quality as a result of rising standards on industry and transportation (although partly offset by rising population and economic activity), successful environmental remediation through federal-provincial collaboration in several key ecosystems (e.g., the Great lakes, the St Lawrence, Georgia Straight) and the rapid growth in renewable energy as a result of government incentives (albeit from a low base). The federal government has introduced a host of policy measures (often in collaboration with the provinces and territories) to make Canada more environmentally sustainable in agriculture, forestry, urban areas, energy efficiency, and waste management.

But is it enough? In 1987, the Brundtland Commission wrote that "the ability to anticipate and prevent environmental damage requires that the ecological dimensions of policy be considered at the same time as the economic, trade, energy, agricultural, and other dimensions."[30] Plainly, this is not fully happening yet in spite of strategic environmental assessments and sustainable development strategies. While decision-makers are more sensitive to environmental issues than ever before, they have not integrated them effectively into departmental strategic plans, performance indicators or internal rewards systems. Partly as a result, overall environmental quality continues to deteriorate in Canada, the number of endangered species rises, fisheries crash, smog remains a problem in populated areas and Canadian emissions of greenhouse gas emissions increase year by year. A 2004 OECD review of Canada's environmental performance concluded that:

- emissions of air pollutants remain high compared to most OECD countries;
- at the current rate of progress, it could take another 20 years or so for Canada to extend and upgrade its water infrastructure to the level required;
- the share of total national area protected is less than the OECD average;
- chemicals of concern are not monitored in the environment on a regular basis.

The leadership deployed to integrate sustainable development into government operations and policies has been episodic. Rhetorical commitments and policy and program changes have struggled to have real impact at the implementation stage. Scientific warnings such as the 2005 Millennium Ecosystem Assessment and the ongoing evidence from the Intergovernmental Panel on Climate Change have not generated much of a sense of urgency. With little sense of urgency, and little leadership, politicians have been easily distracted by national unity, the fiscal deficit, and health care spending issues – all admittedly urgent issues but not ones that address the root causes of unsustainability. SD as an alternative paradigm that enhances environmental integrity, economic prosperity and social justice has struggled to shape decisions of the political and bureaucratic leadership in the federal government. The words and the actions of politicians in this area reveal a state of cognitive dissonance. Politicians know that Canadians have to change their attitudes and behaviour yet they appear unable to sustain the serious modifications to government policies and instruments to achieve these.

One need only look at the rhetorical flourish of the Conservative's Green Plan, the Liberal's Guide to Green Government, and departmental SDS to see that at one level there is a cogent understanding of the nature of the challenge and the quality of the opportunity provided by SD. Yet politicians have been slow to institutionalize the practices that might change outcomes. An example is the Cabinet Directive on SEA which was first introduced in 1990 to govern cabinet and departmental deliberations. It largely fell into disuse and had to be reenergized in 1999. Because it is shrouded in the convention of cabinet secrecy it is still unclear how effective it is in its new iteration.[31] To make progress the sustainability agenda has to be on the prime minister's list of priorities. Under Mulroney, the period around the Green Plan and the Earth Summit was a time of energy and creativity. The early years of the Chrétien government, when implementing the Red Book commitments and the Guide to Green Government was a priority, was also an innovative period. But otherwise, the traditional regional development pressures and the determination of ministers from industry based departments to defend the traditional interests of sectors have led to significant feuding between departments and around the cabinet table (viz., the protracted reform of the *Canadian Environmental Protection Act* and the deadlock on climate change policy). This has sapped creativity and innovation. Hence, during this period at least, traditional concerns aligned to the resource management paradigm have re-emerged to trump sustainable development concerns for the most part.[32] Prime ministers appear to have been reluctant to use the PCO, the prime minister's department, to reign in feuding departments or to hold the deputy minister community accountable for institutionalizing in the bone marrow of the public service the practices politicians introduced with flourish. The intermittent impact of deputy ministers coordinating committees on SD to overcome departmental divisions is a consequence.

Politicians have sometimes overtly blamed the intransigence of the public service as a barrier to making serious breakthroughs. Lucien Bouchard squarely blamed the bureaucracy for the struggles he faced in introducing the Green Plan when he stated "Finance was against it you know. The bureaucrats were against it. Bureaucrats are the worst enemy, I discovered that."[33] Len Good, the DM of Environment Canada, was penalized by the bureaucratic leadership for his efforts to move the Green Plan ahead against bureaucratic opposition. In the recent debate over the decision by the Chrétien government to locate the CESD within the Office of the Auditor General, the Environment minister at the time, Sheila Copps, argued that virulent bureaucratic opposition to an independent Commissioner was a major factor in the government's decision. The CESD has not been able to reach its full potential as a commissioner trapped as it is in an audit office. The CESD has never been able to champion SD the way that the commissioner of Official Languages is expected to champion bilingualism. The structural contradiction of not being able to promote policy and program innovations that might advance SD when this is precisely what is needed at this point in history, because of historic norms and legitimate constraints of the OAG, has become obvious.

While politicians and governments come and go in a democratic political system, the public service is the custodian of long term problems. How does one explain the manifest underperformance of the public service to implement the programs and commitments they undertake? *Our Common Future* provided an insight:

The objective of sustainable development and the integrated nature of global environment/development challenges pose problems for institutions ... that were established on the basis on narrow preoccupations and compartmentalized concerns ... The challenges are both interdependent and integrated ... Yet most of the institutions facing those challenges tend to be independent, fragmented, working to relatively narrow mandates with closed decision processes. Those responsible for managing natural resources and protecting the environment are institutionally separated from those responsible for managing the economy. The real world of interlocked economic and ecological systems will not change; the policies and institutions concerned must.[34]

The CESD has over the past 11 years delineated annually the "implementation gap" between what departments said they would do and what they actually did. Some of this may just be the outcome of a ritualistic policy dance in which bureaucrats never quite believed that the politicians meant what they said and therefore expended the minimal amount of effort to meet minimal expectations. Indeed, many analysts have identified the lack of serious senior level bureaucratic support as an explanation for weak working level commitment to the various new instruments of SD.[35]

Some bureaucrats may be inherently hostile to the paradigmatic shift embodied in sd and do their best to undermine the types of changes that would give it traction. For example, retired Deputy Ministers Ian Clark and Harry Swain dismissed sd as a "utopian framework" entailing "surreal management requirements."[36] An sd approach would constrain bureaucratic flexibility and raise performance expectations. Therefore it should be opposed because "novelty does not last" and "reality eventually prevails." From this perspective managers should not worry too much about performance audits revealing performance failures because the system won't judge them harshly. Given such overt contempt for the realignment of bureaucratic power relations called for by sd within the attitudes of some of the system's leadership it is not surprising that the cesd is able to document policy and program failure so routinely ... including the failure of various iterations of the dm's coordinating committee on sd to make a difference.

While other reasons (including provincial opposition, public opinion, and competing priorities) also contributed, these political and bureaucratic impulses go some way to explaining why Canada has been an international laggard in using economic instruments to pursue higher levels of performance. Essentially, every introduction of an economic instrument had to overcome suspicion and opposition from within the Department of Finance. The result of Canada's slow uptake of economic instruments relative to other countries is that our approach to addressing environmental problems is not as cost-effective as it could be. Consequently, entering the third decade of the post-Brundtland Report era, we have significant policy failures on our hands. If Canada's quality of life remains high, it is likely as much a result of its generous resource endowment, large geography and low population density as concerted government policies. The issue here is not whether progress has been made – it has – but whether it is fast enough given the global and local pressures on natural resources and human health. The fact that this assessment is necessarily subjective speaks to Canada's failure to develop metrics to measure progress.

LOOKING FORWARD: 2007–2027

Can we move nations in the direction of sustainability? Such a move would be a modification of society comparable in scale to only two other changes: the Agricultural Revolution of the late Neolithic, and the Industrial Revolution of the past two centuries. These revolutions were gradual, spontaneous, and largely unconscious. This one will have to be a fully conscious operation, guided by the best foresight science can provide. If we actually do it, the undertaking will be absolutely unique in humanity's stay on earth.

William Ruckelshaus, former director us epa and member wced[37]

Over the past 20 years the sustainability 'revolution' has spawned a flurry of debate and action domestically and internationally. Like all societal transformations, this has been a period of turmoil, creativity and uneven progress.

Traditional assumptions and practices have been challenged. Various drivers of change in the direction Ruckelshaus envisions have confronted legacy barriers of the industrial revolution and have battled it out with varying degrees of success in the energy, mining, forestry, auto, chemicals, urban design, building products, and construction sectors. These battles in the economic sectors have been paralleled within governments, where vacillations in political leadership and bureaucratic intransigence have been major barriers. What "fully conscious" policies, programs, initiatives, or developments are required in the federal government if it is to help advance Canada on the sustainability trajectory?

Political Will and Policy Certainty

Sustainable development is not a Conservative, Liberal, NDP, Bloc Quebecois, or Green idea: indeed, a federal Conservative government developed the 1990 Green Plan, a provincial NDP government introduced the first sustainable development act in Canada (Manitoba) and the federal Liberals made the preparation of departmental sustainable development strategies compulsory. Ruckelshaus was a Republican, not the party most recently associated with environmental protection in the United States. Whichever party is in power has to close the gap between what governments have to do to help Canadians address long term sustainability issues and what the federal government actually does. Like the Mulroney and Chrétien governments before it, the Harper government is currently experiencing a degree of cognitive dissonance between what it knows are the expectations of Canadians for innovative solutions to address climate change and other SD issues, and what it has delivered to date. The first Harper minority government between 2006 and 2008 experienced a steep learning curve and significantly modified its views on environment and SD issues (e.g., climate change). During this period, it largely governed as an opposition party temporarily in power running against the record of its predecessors.[38] It has to start acting like a government if it wants to convince Canadians that it has the vision and capability to be trusted to lead Canada into the 21st Century. Its inability to win a majority government in the 2008 federal election and subsequent developments leaves this an open question. Yet, as *Our Common Future* noted, "in the final analysis, sustainable development must rest on political will."[39]

The Conservatives would be well advised to embrace the expert SD community and build on their ideas and experience in developing policies for emission reductions, infrastructure improvements, urban redesign, and resource management. The Canadian Council of Chief Executives has argued that emissions reductions and a smaller ecological footprint are entirely consistent with improved economic performance. Interestingly, the CCCE states that the first thing that is needed is a national plan of action that includes government, industry, and consumers. The CCCE goes on to argue that a major key to success is creating "price signals" to encourage sustainable use of energy through the use of "emissions trading" and "environmental taxation."[40] The NRTEE has been making similar arguments.

However they define sustainability, experts agree that the changes required to achieve it will take time. This is not only because major investments need to be made to make our transportation and built infrastructure more energy-efficient, and these take time, but also because sustainability implies a gradual but significant transformation in our production and consumption patterns as well. The depth of the changes required makes it important to achieve policy stability so that investors have an incentive to make the long-term commitments needed. This will require the development of a political consensus or a collective vision of the goals we are seeking to achieve. For example, the Harper Conservative government and the Dalton McGuinty led Ontario Liberal government agree that financial support packages to the North American auto manufacturers in 2009 and beyond are contingent upon the "Big Three" dramatically enhancing the fuel efficiency of the vehicles they produce. While opinion polls suggest that Canadians expect more leadership from their governments on environmental issues, they remain divided over both the direction and urgency of change. Federal political leadership is thus required not only to project a vision of a sustainable future for Canada but also in putting in place a believable and stable policy climate to reach that future that will provide incentives to both individuals and investors.

Coordination within the Federal Government

There are several non-partisan institutional modifications that any federal government could institute in the short term and that subsequent governments could retain and strengthen over the next 20 years around the cabinet decision-making process and the coordination of interdepartmental performance. An important first step was achieved in June 2008 when the *Federal Sustainable Development Act* (FSDA), which emerged from a minority Parliament, received Royal Assent.

The FSDA addresses a number of the shortcomings discussed above which were revealed by the previous decade of institutional experimentation. The peculiar Canadian condition of having a sea of departmental SD strategies but no overarching government strategy has been addressed with the legal requirement to produce and update a federal strategy every three years. "The Federal Sustainable Development Strategy shall set out federal sustainable development goals and targets and an implementation strategy for meeting each target and identify the minister responsible for meeting each target."[41] This is an important advance because the paucity of serious goals and targets and the lack of ministerial accountability had been a major weakness of the previous approach. Regarding accountability, the FSDA requires that "Performance-based contracts with the Government of Canada shall include provisions for meeting the applicable targets referred to in the Federal Sustainable Development Strategy and the Departmental Sustainable Development Strategies."[42] This should help address the problem of holding departmental senior officials responsible for the performance of their departments, which had been a fatal weakness. Additional clauses requiring departmental SD strategies to contain objectives and plans that comply with and contribute to the overall federal SD strategy and the power to make regulations prescribing the form in which departmental strategies are to be prepared and the information required to be contained in them should help force greater interdepartmental cooperation on policies and overcome some of the coordination problems confronted by previous attempts to green government and influence procurement, for example.

The FSDA creates three new institutions. A sort of 'cabinet committee' styled "A committee of the Queen's Privy Council for Canada … shall have oversight of the development and implementation of the Federal SD Strategy."[43] Rather than create a SD office in the PCO, the FSDA creates this office within Environment Canada to "develop and maintain systems and procedures to monitor progress on implementation of the FSDS."[44] An SD Advisory Council will be chaired by the minister of the Environment and composed of a representative from each province and territory, and three representatives each from aboriginal, environmental, business, and labour organizations. While these outcomes are compromises negotiated amongst the parliamentary parties, they all respond to problems identified by analysts as problematic in the earlier approaches.

As is always the case in public policy, the actual implementation of the Act will determine the effectiveness of the new approach. There are still outstanding institutional needs. For example, the Privy Council Office needs to strengthen departmental compliance with the cabinet directive on SEA and Treasury Board Secretariat must strengthen system-wide accountability and reporting requirements. The speaker struck down on a technicality the proposal in the original Bill to make the CESD an independent Officer of Parliament. The lack of an independent CESD remains a weakness in the Canadian system and a renewed mandate for an independent Commissioner awaits a future government.

Ecological Fiscal Reform

One priority stands out over all others in considering what Canada needs to do over the next twenty years. This priority is harnessing market forces to ensure that prices better reflect environmental externalities. Canada needs to make much greater use of economic instruments in order to promote sustainable development. Canadians live in a market-based society. Many of the decisions they make as consumers, investors and lenders are based on market prices, most of which do not reflect environmental costs. This leads to a large scale misallocation of societal resources. Renewable forms of energy have difficulty competing with fossil fuels because the latter's prices do not incorporate a premium for climate change. Public transit finds it difficult to compete with automobile commuting when governments subsidize road transport to a greater extent than public transport. Organic food is in part more expensive than the products of industrial agriculture because the latter do not pay a penalty for polluting waterways with pesticides. Disposable products are cheap in part because the costs of their disposal is hidden in municipal taxes rather than incorporated to the product price. There are significant limitations to the degree to which regulation can overcome this market dynamic. While regulation is required to ensure the baseline protection of basic ecosystem conditions (clean water, clean air, etc.), only economic instruments can address the underlying production and consumption challenges we face in a way that stimulates and builds on the innovation and economic development dynamics of the market place.[45]

Canada should draw on the growing international experience with economic instruments. Although there are various legal, geographical and other considerations that may limit the direct applicability of some of the foreign experience, there are important examples of economic instruments that have been demonstrated to be effective in addressing an issue of importance to Canadians and that could be implemented in Canada without significant law or institutional reforms.[46] International experience also provides many valuable process and design lessons for Canadian efforts to introduce economic instruments.

Many European countries are moving beyond the piecemeal approach of using economic instruments for discrete issues and towards more fundamental ecological fiscal reform. This trend holds prospects for significantly changing the way those countries promote sustainable development. Using increased revenue from consumption taxes to reduce distortionary taxes on capital or labour offers the long-term prospect of enhancing the efficiency and pervasiveness of incentives for environmentally sound decision-making while stimulating economic development and employment.

British Columbia's green tax shift, the federal Liberals embrace of this idea in the 2008 election campaign, the federal Conservative government's and several province's endorsement of a "cap and trade" regime for reducing GHG from industry, suggests that in 2008, after twenty-five years of debate, the idea of using economic instruments to achieve environmental objectives had finally entered the mainstream of political discourse. Ironically, the extraordinary run-up in world oil prices that occurred over 2007 and 2008 started to have the same effect on the energy markets that the much-maligned carbon taxes were designed to have: efforts to increase energy efficiency and develop renewable forms of energy were given a new impetus. The equally dramatic drop in oil prices in the fall of 2008 and the poor electoral performance of the Liberals in the 2008 election underscores that there is still plenty of fluidity in the debate on the use of economic instruments. Nonetheless, the idea that carbon needs to be priced has entered the mainstream and must be now be implemented in policies and markets.

Regardless of the precise direction Canada takes, the main lesson from the foreign measures is the importance of developing domestic experience with what works and what does not (both in terms of the effectiveness of certain instruments as well as in the manner in which they are introduced). Experience is essential to overcome scepticism and to ensure appropriate design. Fundamentally, such experience can only be gained through implementation, ongoing innovation and careful scrutiny of outcomes and lessons learned.

Information

Governments started to manage the economy once they developed a comprehensive system of national accounts showing in detail what changes were occurring in the economy. On the basis of these accounts, it became possible to model the impact of fiscal and monetary policy and to design measures to counteract economic cycles or boost productivity. While economic indicators do not provide an exact analog for environmental indicators, they do underline the truism that "one manages what one measures."

Canada needs better information about its environment including about the impact of federal policies. As a country, we need to define "how clean is clean?" so that we can design appropriate environmental policies at an acceptable cost. Some countries – Sweden is currently the best example – have done so. Within an overall goal of becoming sustainable within a generation, they have set a limited number of goals and measurable targets against which to evaluate progress. An Environmental Objectives Council reports annually to Parliament. Canada should do the same. Indeed, this could be a key role for a revamped CESD.

Sweden has found that the major benefit of these objectives has been to focus everybody's attention (all levels of government, industry, NGOs) on the highest priorities, leading to greater efficiency and coherence in environmental protection efforts in the country because they replaced a large patchwork of different standards with different schedules for implementation. They create a transparent and stable policy framework for all governmental programs (i.e., sectoral departments cannot pursue policies that would undermine the achievement of the objectives). They are supported by 71 more specific interim targets (also set by Parliament) and by objectives set at the regional and local levels. While each goal provides the overall vision of the desired environmental state for a particular issue, most of the targets are quantitative and provide indicators to measure progress. National environmental objectives in Canada would help rationalize the current fragmented departmental sustainable development strategies which yield thousands of commitments to little effect because they scatter rather than focus resources. Canada took a first step towards such national objectives when the 2008 *FDSA* mandated the establishment of federal sustainable development targets by 2010. Although these targets are expected to be developed through a less-inclusive process than the Swedish ones and to be less ambitious, they too may help to create a more coherent policy framework.

In a related but different vein, the federal government also needs to build a more solid foundation for policy formulation. In the first instance, this will require improving current tools, such as *SEA*, and broadening the application of integrated metrics such as Genuine Progress Indices. But it will also require increasing the policy capacity of civil society to address these issues. Contrary to policy fields such as health or education or social welfare, the sustainable development infrastructure for policy analysis and debate is very thin. Few independent voices exist who can speak authoritatively in these areas and overall public awareness remains low. The societal response to sustainable development will require major adjustments across society, however, and democracies find it difficult to force their members to change their lifestyles. It is essential therefore to continue to raise public awareness about the serious and long term nature of sustainable development issues, such as climate change, and the hard choices modern societies must make, particularly with respect to energy use. This understanding is critical, because it underpins any political consensus to take determined action. As well as raising awareness, the government should also assist Canadian institutions in all societal sectors to develop their own capacities for analysis, innovation, and interaction with other relevant stakeholders.

CONCLUSION

More effective government coordination and leadership, a renewed mandate for an independent CESD, ecological fiscal reform, a stronger information base, a set of explicit environmental objectives and a more solid foundation for policy formulation are necessary but insufficient reforms to ensure that the federal government promotes sustainable development over the next twenty years. It also needs to pursue substantive policies in areas such as climate change, urban transportation, nature protection, the regulation of toxic chemicals and water to move Canada to a more sustainable path. We believe that these policies will need to be supported by the fundamental institutional and instrument design reforms we have outlined above.

NOTES

1 World Commission on Environment and Development (WCED), *Our Common Future* (Oxford University Press, 1987), 9.

2 Ibid., 312.

3 G. Bruce Doern and Thomas Conway, *The Greening of Canada: Federal Institutions and Decisions* (University of Toronto Press, 1994).

4 Brian Mulroney, *Memoires* 1939–1993 (Douglas Gibson Books, 2007); Toby Heaps, "Brian Mulroney: No regrets," *Corporate Knights* (Summer 2005), 31–2.

5 National Task Force Environment and Economy, *Report to the Canadian Council of Resource and Environment Ministers* (1987), 11; A. Dale, C. Ling, and C. Spencer, *The National Round Table on the Environment and the Economy: Expanded Decision-making for Sustainable Development* (Royal Roads University, 2006).

6 William Lafferty and James Meadowcroft, eds. *Implementing Sustainable Development: Strategies and Initiatives in High Consumption Societies* (Oxford University Press, 2000); Barry Dalal-Clayton, *Getting To Grips with Green Plans: National Level Experience in Industrialized Countries* (Earthscan, 1996).

7 Canada, *Canada's Green Plan – for a healthy environment* (1990), 4–6.

8 Glen Toner, "The Green Plan: From Great Expectations to Eco-Backtracking to Revitalization?" in Susan Phillips, ed. *How Ottawa Spends 1994–95: Making Change* (Carleton University Press, 1994).

9 Dalal-Clayton, *Getting to Grips* (1996).

10 Luc Juillet and Glen Toner, "From Great Leaps to Baby Steps: Environment and Sustainable Development Policy Under the Liberals," in Gene Swimmer, ed. *How Ottawa Spends 1997–98: Seeing Red – A Liberal Report Card* (Carleton University Press, 1997).

11 An Environment Committee had existed before.

12 Canada, *Guide to Green Government* (1995), 1.

13 For a more detailed assessment of this innovation see Francois Bregha, "The Federal Sustainable Development Strategy: Why Incrementalism Is Not Enough," in G. Bruce Doern, ed. *Innovation, Science, Environment: Canadian Policies and Performance, 2006–2007* (McGill-Queen's University Press, 2006); Francois Bregha, "Missing the Opportunity: A Decade of Sustainable Development Strategies," in Glen Toner, ed. *Innovation, Science, Environment: Canadian Policies and Performance, 2008–2009* (McGill-Queen's University Press, 2008); Glen Toner and Carey Frey, "Governance for Sustainable Development: Next Stage Institutional and Policy Innovations," in G. Bruce Doern, ed. *How Ottawa Spends 2004–05: Mandate Change in the Paul Martin Era* (McGill-Queen's University Press, 2004).

14 M. Winfield, M. Anielski, H. Benvides, and A Kranj, *Governance Tools for Sustainable Development within the Government of Canada* (Pembina Institute, 2002); Stratos, *Governance Models for Sustainable Development* (2002).

15 House of Commons Standing Committee on Environment and Sustainable Development (SCESD), *Report on the Commissioner of the Environment and Sustainable Development* (1994).

16 SCESD, *Hearings on the Commissioner of Environment and Sustainable Development* (2007).

17 National Roundtable on the Environment and the Economy (NRTEE), *Advice on a Long-term Strategy on Energy and Climate Change* (2006); *Interim Report to the Minister of the Environment* (2007).

18 Jack Mintz, *2007 Tax Competitiveness Report: A Call for Comprehensive Tax Reform* (C.D. Howe Institute, 2007). Available at: www.cdhowe.org/pdf/ commentary_254.pdf; Jeffrey Simpson, Mark Jaccard, and Nic Rivers, *Hot Air: Meeting Canada's Climate Change Challenge* (Douglas Gibson Books, 2007).

19 Kernaghan Webb, ed. *Voluntary Codes: Private Governance, Public Interest and Innovation* (Carleton University, 2004).

20 Robert Gibson, ed. *Voluntary Initiatives: The New Politics of Corporate Greening* (Broadview Press, 1999).

21 Organization for Economic Cooperation and Development (OECD), *Institutionalizing Sustainable Development* (2007).

22 OECD, *Performance Review – Canada* (2004).

23 Glen Toner, ed. *Sustainable Production: Building Canadian Capacity* (University of British Columbia Press, 2006).

24 Stratos, "Economic Instruments for Environmental Protection and Conservation: Lessons for Canada, for the External Advisory Committee on Smart Regulations" (Privy Council Office, 2003).

25 Commissioner of Environment and Sustainable Development (CESD), *Report of the Commissioner of Environment and Sustainable Development to the House of Commons* (2004).

26 CESD, *Report of the Commissioner* (2005 and 2008).

27 CESD, *Report of the Commissioner* (2005); Bregha, "The Federal Sustainable Development Strategy" (2006); Bregha, "Missing the Opportunity" (2008).

28 NRTEE, *Environment and Sustainable Development Indicators for Canada* (2003).

29 Stratos, *Enhancing Public Access to Corporate Environmental Performance Information* (2002).

30 WCED, *Our Common Future*, 10.

31 The Canadian Environmental Assessment Agency was leading an evaluation of the Cabinet Directive at the time of writing.

32 Glen Toner, "Canada: From Early Frontrunner to Plodding Anchorman" in William Lafferty and James Meadowcroft, eds. *Implementing Sustainable Development: Strategies and Initiatives in High Consumption Societies* (Oxford University Press, 2000).

33 "Bouchard Laments Green Plan Fate," *Montreal Gazette*, 8 April 1991, A7.

34 WCED, *Our Common Future*, 9.

35 CESD, *Report of the Commissioner* (2005); Bregha, "The Federal Sustainable Development Strategy" (2006).

36 Ian D. Clark and Harry Swain, "Distinguishing the real from the surreal in management reform: Suggestions for beleaguered administrators in the government of Canada," *Canadian Public Administration* 48.4 (Winter 2005), 453–76.

37 As quoted in David R. Boyd, *Sustainability Within a Generation: A New Vision for Canada* (David Suzuki Foundation, 2004). Available at: www.davidsuzuki.org/files/WOL/DSF-GG-En-Final.pdf.

38 Glen Toner, "The Harper Minority Government and ISE" in Glen Toner, ed. *Innovation Science, Environment: Canadian Policies and Performance 2008–2009* (McGill-Queen's University Press, 2008).

39 WCED, *Our Common Future*, 9.

40 Canadian Council of Chief Executives (CCCE), *Clean Growth: Building a Canadian Environmental Superpower* (2007). Available at: www.ceocouncil.ca.

41 Canada, *Federal Sustainable Development Act (FSDA)*, Clause 9:2.

42 Ibid., Clause 12.

43 Ibid., Clause 6.

44 Ibid., Clause 7:1.

45 Mintz, *2007 Tax Competitiveness Report* (2007).

46 Stratos, "Economic Instruments for Environmental Protection and Conservation" (2003).

4 Post-Brundtland 2007: Governance for Sustainable Development as if It Mattered

ANN DALE

This chapter is dedicated to Dr. Paz Butterdahl, head of the Human Security and Peacebuilding Program, Royal Roads University, who never failed to speak out when others were afraid. It is indeed true that inside every Chilean, there is a general.

8 April 1941 – 8 October 2007

As the 21st century progresses, the inability of our outmoded governmental institutions to address the deeply complex ecological, economic, and social challenges that we face is increasingly apparent. In 1987, the Brundtland Commission warned that:

The integrated and interdependent nature of the new challenges and issues contrasts sharply with the nature of the institutions that exist today. These institutions tend to be independent, fragmented, and working to narrow mandates with closed decision processes. Those responsible for managing natural resources and protecting the environment are institutionally separated from those responsible for managing the economy. The real world of interlocked, economic, and ecological systems will not change; the policies and institutions concerned must.[1]

The Commission subsequently recommended several steps for institutional and legal change in the basic ways we govern ourselves. However, since this warning was issued our method of governing in Canada has changed little, and the problems, if anything, have grown even more complex.

Many experts have identified that one of the major barriers to the implementation of sustainable community development is governance.[2] Others have referenced fundamental disconnections such as those between federal, regional, and local governments, between rural and urban communities, and critically, between the business and research communities.[3] As one example, provincial and federal policy for environmental protection is undermined by municipal development plans that sprawl into farmlands and green space.[4] Such sprawl radically increases the need for private automobiles, challenging the realization of federal climate change commitments. Such examples can be found at all levels of government; we are far too often working at cross purposes.

Governance[5] can be defined as the sum of many ways individuals and institutions, public and private, manage their common affairs. It is a continuing process through which conflicting or diverse interests may be accommodated, and cooperative actions may be taken. It includes formal institutions and regimes empowered to enforce compliance, as well as informal arrangements that people and institutions have agreed to as perceived to be in their interest.[6] In this chapter I focus specifically on formal institutions, particularly those at the federal level. I have observed federal governance from several perspectives; as a federal government civil servant for 23 years, as a full-time researcher and professor studying government's role in sustainable development, and as a facilitator of several policy-relevant research projects.[7]

What was clear from these experiences is that there are fundamental disconnects in the way we govern with respect to our relationship with the natural world. This is particularly troubling given that there is overwhelming evidence that our ecological capital is continuously declining, and many argue that our economic and social progress has been at the expense of this decline.[8]

Institutional failure to address these disconnections, and our institutional incapacity to bridge the solitudes, silos, and stovepipes[9] that are antithetical to the resolution of broad horizontal issues such as climate change and biodiversity conservation, are major barriers to any significant achievement of sustainable development by Canadian communities.[10] Our current governance system seems incapable of going back to first principles, that is, re-evaluating the original context, critically analyzing institutional change independent of vested interests, traditional power and brokerage relationships, particularly federal/provincial relationships, and revising or changing previous decisions – what a colleague has referred to as the "limits to democracy."[11]

Both the problems and the solutions rest with the form of our institutions. Institutions are defined as "organized sets or collections of human, informational and material resources, available for collective action and providing decision rules and processes for adjusting and accommodating over time conflicting demands . . . of groups and individuals in society."[12] With respect to our response to Brundtland, the question we need to ask is whether Canadian federal institutions have failed to respond to the sustainable development imperative? After talking about the specific institutional response by the federal government – namely: institutional capacity for horizontality; strategic policy analytical capacity; policy coherence and policy alignment between governments; integrated decision-making; expanded decision-making processes, and political will and leadership – I will then discuss the broader governance implications.

A FAILURE TO RESPOND?

The answer to the question asked above is affirmative, for many reasons. First, the perpetuation of an inflexible, hierarchical, decision-making system works against horizontality. Sustainable development is a broad horizontal imperative; its issues are inherently interdisciplinary. Second, the lack of a systems approach to increasingly complex challenges reinforces piece-meal ad hoc policy development. Third, the established vertical departmental silos are antithetical to responding to the resolution of complex socio-ecological problems. Interdisciplinarity implies there is some common conceptual or systemic framework that underpins the research and policy framework(s). It requires the conscious searching for unifying concepts that foster and reinforce understanding across disciplines, between government, industry, and the research community. The question is whether our institutions, our government departments, and agencies, are even designed to do this.

Funtowicz and Raven argue that a post-normal scientific approach is also necessary, that is, a plural "systemism" in which both the parts and the whole – analysis and synthesis – are essential elements.[13] In addition, sustainable development issues, such as climate change and biodiversity conservation, involve conditions of high variability and complex interactions.[14] Put differently, given the number of sectors and multiplicity of actors that must necessarily be involved in any potential solutions, the contextual nature of sustainable development requires a multi-stakeholder approach to decision-making well beyond the current capacity of federal government departments as we now know them. It requires an expanded decision-making context.

The issues of sustainable development are also highly normative; they involve complex issues of polity and culture. As well, the type and nature of information required to discuss and to evaluate such issues as climate change and biodiversity for decision-making is profoundly different than most other policy issues. These issues require novel information about the interaction between human activity systems and natural systems, and often across generations. Accordingly, the information required necessarily involves large and long-term future spatial scales, as well as intergenerational transfers. This long time horizon coupled with the spatial dimensions of many ecological processes, combined with long lags in economic and social change, means that the spatial and temporal boundaries of sustainable development problems are inherently more complicated than the issues currently confronting governments. Resolution of sustainable development issues also requires the ability to recognize and respond to negative feedback information from natural and human systems, demands broad institutional co-ordination, as well as flexible and adaptive capacity.[15]

There are deep structural reasons for the failure to implement sustainable development, both at the federal institutional level, addressed below, and the municipal level. The latter was primarily built around a 1950s model of development that is clearly no longer viable. Built on a model of high growth, and a highly decentralized suburban model around individual car use and single-use buildings, it is beholden to development interests, and continues to be one of the most unsustainable forms of the built environment, especially given current climate change projections into the future.

GOVERNANCE IMPLICATIONS

Although much has been written and discussed about sustainable development in the academy, it did not realize popular acceptance and diffusion worldwide until the publication of *Our Common Future* in 1987. Although hundreds of definitions now exist,[16] there is still debate over defining the concept and arguments persist about the need for a national sustainable development strategy in Canada, as well as a common framework. Often, definitional arguments can be used to avoid changing, as is the argument that more research is needed before being able to act; for example one needs to question why the majority conclusions of the world's scientists on the Intergovernmental Panel on Climate Change has taken over a decade to reach the political agenda in an age of mass communications. Frequently, framework debates hide old power and conflicts often seen as irreconcilable. While disagreement exists among different communities about the usefulness of the concept of sustainable development and which framework is preferable,[17] and there has been much castigation of the term as an oxymoron; many others believe the usefulness of the term lies in its constructive ambiguity. It has brought people together at round tables who have traditionally been adversaries, and it has essentially linked the earlier separate concepts of sustainability and development. Thus, its strength lies in its ability to bridge the traditional left-right polarization that has characterized much of the debate in this country since the 1970s, in its constructive ambiguity.

Turning definition into practice, however, has been difficult. Sustainable development has encountered several points of failure. These include a continual resolution of multiple and often diametrically opposed value systems; use of a necessarily diverse range of expertise, both social and natural science; solutions that are beyond the ability of any one sector or level of government to address or implement; solutions that most often will be beyond the capacity of any single organization to address; and thus, the demand for novel relationships among researchers, governments, business, the scientific community, and civil society. In short if we are to turn theory into practice, we must overcome a formidable implementation gap.

INSTITUTIONAL CAPACITY
FOR HORIZONTALITY

Canada has a rich history of both theoretical and round table discourse on sustainable development, yet this discourse has neither deeply penetrated

Canadian institutions nor Canadian communities. Many communities have already experienced reviews, produced consultant reports on what is needed for change, and yet, they have never been implemented. One of the major reasons for this implementation gap is the gridlock in the planning and implementation processes for decision-making all Canadian communities face. This gridlock is not because of lack of research, knowledge, and information residing in communities, but rather has arisen because of the solitudes, silos, and stovepipes that characterize the research, business, and government sectors. It is multi-faceted and involves, among other things, a lack of coherent dialogue, congruence between political levels, political will and a 'sustainable development' ethos among various government levels and community stakeholders. Further contributing to these failures are critical information gaps and large coordination failures about channeling the proper resources to the right target[18].

HORIZONTAL STRUCTURES CANCELLED

Ministry of Urban Affairs (1980)

Ministry of State for Social Development (1985)

Ministry of State for Economic Development (1985)

Green Plan (1992–1993)

Tri-Council Eco-Research Program (1991–1995)

Science Council of Canada (1990)

Economic Council of Canada (1990)

Canadian Environment Advisory Council (CEAC) 1990

Provincial Round Tables (various dates)

Institute for Sustainable Development Research (2002)

State of the Environment Reporting (1995)

It is important to understand how solitudes, silos and stovepipes work in the government context. The term, *solitudes*, refers to deep cleavages based on language, cultural, religious, and other forms of exclusion. *Silos* describe the divisions between sectors such as research and business, researchers and government policy-makers, and between governments. *Stovepipes* are separations within an organization or business, for example, between and within government departments and in universities, among disciplines. The misplaced concreteness[19] of these stovepipes is a major barrier to any potential for effective government policy congruence within government departments and policy alignment between levels of government. It is clear simply by viewing the current federal government structure that there has not been needed structural reform from our history of hewers of wood and drawers of water – particularly with the continuing and concretized silos between Environment and Finance; between Environment and Natural Resources, between Fisheries, and Oceans, and Agriculture and Agri-Foods Canada. Klein further describes the dilemma:

Modern societies are increasingly dominated by the unwanted side effects of their differentiated sub-systems, such as the economy, politics, law, media, and science. These systems have developed their own running codes, to use Niklas Luhmann's term that enable them to be highly productive. However, differentiation produces imminent side effects in other fields that cannot be handled within the codes of the system. Indicative of this development, the problems of society are increasingly complex and interdependent. They are not isolated to particular sectors or disciplines, and they are not predictable. They are emergent phenomena with nonlinear dynamics. Effects have positive and negative feedback to causes, uncertainties continue to arise, and unexpected results occur.[20]

Is there then any evidence of a reorganization of federal government department silos and stovepipes, or a shift to ecosystem-based management,[21] or adaptive management?[22] If so, it is very slight and halting. One of the major barriers for governments interested in basic institutional reform is that governments have inherited from the nineteenth century a way of thinking and organization that is structured around old problems, not modern day issues. For example, nineteenth and even twentieth century thought organized government almost exclusively around themes of National Defence, Transportation, National Resources, Security, the Environment, Employment, and so forth. But, the twenty-first century demands quite different responses to the critical public policy issues now facing Canadians. Climate change, as only one example, demands new ways of aligning municipal, provincial, and federal government policies, incentives, norms and standards, as well as critical infrastructure to both anticipate and mitigate human impacts. It is only when governments have the capacity to organize dynamically around current and emerging issues and proactively respond to their modern context that they become responsive to their citizenry, and remove the "limits to democracy."

Sustainable development issues require totally different ways of government organizing – organic, self-organizing and more flexible structures that respond to the different contexts surrounding each issue. Sustainable development issues are beyond the capacity of any single jurisdiction and beyond traditional problem-based government structures to solve. They are adversely affected by piece-meal and uncoordinated policy systems that lack the support of clear government priorities. The problems here are as basic as questions of democratic legitimacy: who defines the problems, who gets to set the priorities, who gets to frame the questions and who gets to decide who the experts are? Sustainable development issues as well are not easily bounded. Indeed, there is often a real mismatch between political and ecosystem boundaries. Put in today's context, environmental issues transcend political boundaries. This demands a move to ecosystem-based management and planning and decision-making, supported by government structures that are organized around specific issues, or domains of appreciation.[23]

Finally, deputy accountability has to change to reflect modern society, an adaptive mix of responsibility for horizontal issues with appropriate accountabilities balanced with vertical operational requirements. By leaving everything vertical, especially policy and new strategic initiatives, there is no capacity for horizontality that is innovative and that would lead to the creation of new networks and reorganized constituencies, especially critical to expanded decision making contexts.

STRATEGIC POLICY CAPACITY

There has been a systemic shrinking of the space in the federal system for analytical policy capacity, never mind strategic policy capacity. There are many reasons, both small and large "p" political, for this shrinking but one reason with major ramifications for the federal system, at first blush seemingly small, has been the augmentation in political staff beginning with the Mulroney administration in the 1980s and the subsequent distrust between the political and bureaucratic levels of government. As political staffs grew, direct access between a minister and their deputy was compromised, and frequently, deputies now meet with senior-level political staff, instead of directly with the minister. Gone are the days when deputies and ministers used to meet every day, and equally gone is the space to inform and educate at a peer level about constraints and barriers, with a critical disconnect opening between internal policy development and political realities. Without critical peer discourse, the opportunity for developing strategic and operational policies that integrate both political and public service imperatives makes for failed discourse and analysis within the federal system.

Without the opportunity to meet physically and directly without small "p" political interference, the opportunity for meaningful, direct, and open dialogue is no longer possible.[24] There is no space for building trust, or social capital, the glue that lubricates professional relationships, and facilitates the taking of risk and innovation, making for a more risk adverse policy development process. With a more risk adverse public service cadre there is less opportunity for building networks of collaboration between levels of government. This is damaging as such networks are evidently even more critical given the horizontality and scale of sustainable development issues. Is there a relationship between this and the lack of grand visions for the development of our country? Would initiatives such as the Granville Island redevelopment and The Forks development in Winnipeg be possible by today's governments?

Paradoxically, political staff now often deal directly with public servants, further blurring the boundaries and necessary separation between political and public service officials. This separation was traditionally seen as necessary to provide critical continuity between changes in political administrations and to ensure the safeguarding of the public good and continuity by a professional public service. This blurring of boundaries has further politicized the policy development process, coupled with the blurring of lines with public service communications functions. This has led to fuzzy accountabilities, a lack of clear responsibilities, and subsequently, a lack of any meaningful public commitments to concrete goals, objectives and targets in public service communications material.

As well, the business community and quasi-government organizations have lobbied for decreased internal analytical capacity, with a subsequent outsourcing of this function to consultants, think tanks, and quasi-governmental agencies, many of who are dependent upon governments for most of their funding. This decrease in internal policy advisory capacity means there is no countervailing force against external lobbying, leaving ministers without a trusted, institutionalized check and balance from external influences and political staffs.

There appears to be a serious lack of strategic policy analytical capacity, attributable partly to the downsizing of the Mulroney administration in the 1990s, ending in 1995. The impact of this direction was that most of the senior policy development community took generous executive buyout packages, leaving a critical skill vacuum and severe gaps in institutional memory. This has ramifications for our ability to anticipate and prevent and to negotiate internationally, for example, trans-boundary water issues.

INTEGRATED DECISION-MAKING

Sustainable development can be defined as the reconciliation of ecological, social and economic imperatives.[25] It is about intermediate and long-term integration and the pursuit of all the requirements for sustainability simultaneously while seeking mutually supportive benefits.[26] Coherently interrelated institutional structures and processes of planning[27] are a critical element of this integration as stated earlier in this chapter. Evidence about the lack of integration was articulated by Jim MacNeill as the forgotten imperative of sustainable development: "we have failed dismally in our attempts to merge environment with economics in our processes of decision-making – in the cabinet chambers of government, in the boardrooms of industry and in the kitchens of our homes."[28]

Integrated decision-making evokes the following characteristics: reconciliation of multiple and often conflicting use imperatives; place-specificity embedded in a comprehensive systems approach; interdisciplinarity and applied science; open policy development processes[29] and expanded decision-making processes. It is decision-making that simultaneously deliberates five distinct capitals – natural, social, human, physical and financial[30] incorporating the following management tools – eco-zone management, adaptive management, and the precautionary and subsidiarity principles.

EXPANDED DECISION-MAKING

Besides interdisciplinarity, sustainable development also requires transdisciplinarity. The latter necessitates interactions not only between disciplines, but also with government, affected populations and between sectors. It also requires a multi-stakeholder approach to decision-making since it is beyond the capacity of one sector, level of government, and profession to solve and more critically to implement.

Since sustainable development is never really achievable,[31] as it inherently involves a dynamic relationship between two complex living systems, the human and the ecological, it requires sustained dialogue in every community about its particular dynamics. Critical to the social change that is necessary for its implementation is a literate, cognizant, and actively engaged civil society. This requires a move from traditional government consultation to dialogues, where governments are also mandated to educate people as well as solicit feedback on existing and new government policies and programs. Fundamental to expanded decision-making contexts are principles of openness, transparency, inclusivity and a commitment to diverse forums for public participation, both face-to-face and virtual. Inclusivity requires deliberative design and strategies for community engagement, it just does not happen, without active planning and strategic identification of the critical stakeholders needed for both knowledge diffusion, as well as implementation. One way to bind such forums is through issue-based expanded policy development, depicted below. As well, the co-construction of public policy allows for social innovation in public policy design and more critically, its implementation.[32]

POLICY COHERENCE AND POLICY ALIGNMENT

Overlapping mandates, duplication and competing jurisdictions further exacerbate gridlock and implementation gaps. In British Columbia, for example, the Fraser River is 'protected' by no less than sixty-two different government agencies.[33] As well, policies, codes, and standards vary enormously across and between governments, and often are simply inconsistent.[34] In addition, initiatives at community levels are often stymied because of a lack of policy alignment between regional, provincial, or even national levels. The result is too often that planning is disconnected from actual implementation and is undertaken without regard for higher level consequences or impacts. Clearly, sustainable development, as a broad, cross-cutting domain of appreciation for all levels of government requires policy integration, along with ameliorated relationships between government and non-government institutions and the creation of longer-term planning horizons in government.[35]

Aside from the absence of longer-term planning, lack of a coherent vision about the meaning of sustainable development and its governance, mitigates against any attempts for policy coherence at the federal level in Canada. There is an urgent need to implement policy coherence within governments, starting at the municipal level, to ensure implementation of Integrated Community Sustainability Plans (ICSPs), followed by policy alignment between levels of government. There is a pressing need to review the existing legal framework as it affects the environment, as the present law is at best a patchwork in dealing with on the ground implementation of sustainable infrastructure and the environment, and varies enormously from province to province and municipality to municipality.[36] Policy alignment between levels of government and congruence within one level of government vertically requires political will and leadership. Without this leadership, at the federal level, cabinet committees remain aligned around the traditional problems, reinforcing vertical departmental organizations, and their vested constituencies of client interests.

POLITICAL WILL AND LEADERSHIP

At the highest levels there is still much to do to set the stage for sustainable development at all levels of operation. Besides, the lack of a national vision and shared meaning about the urgency of key issues, Canada is the only Group of 8 (G8) country that does not have a national transportation strategy. Neither do we have a federal water security strategy, a climate change adaptation and mitigation strategy, nor an energy security strategy. At a specific policy level, although we have had a federal environmental assessment process in place since 1990, it has never adequately applied, as governance is lacking. Audit reviews of 2004 and 2005 reveal that left to each individual ministry, the policy is inconsistently interpreted and applied. Although a legal requirement, it has never been assigned the appropriate accountabilities.

As previously mentioned, the 19th century definition of departments and mandates defined the primary interests of that time, high growth, and development of the country, with no concern for the environment. Since then, other mandates have been appended to the primary mandate, but the primary always takes precedence, and there is never any return to first principles and systemic integration of the 'add-ons' into the policy development process. This 19th century institutional structure is also a formidable barrier to our six large urban centres and municipalities who are attempting to better align urban financial and other capacities with their economic, social, and environmental responsibilities.

The leadership role of government would change from a command and control model to becoming a leader of collaborative processes among diverse actors that produce integrative or holistic understandings of challenge and potential solutions.[37] Governments are the most honest conveners of such collaborations given their mediating role in democratic society. If it is left to the private-public sector to lead, governments risk becoming increasingly marginalized.

A NETWORKED GOVERNANCE MODEL

It is clear that 21st century ecological imperatives necessitate a fundamental restructuring of governments and the corresponding cultural and organizational shifts needed to address critical issues of local vulnerability, flexibility, and subsidiarity.[38] Given the formidable barriers articulated above, we can no longer afford to maintain existing governance arrangements, while waiting for this overall restructuring, rather, it may be prudent to use existing structures in new ways, employing both horizontality for policy development and verticality for operational imperatives taking into consideration the characteristics of sustainable development and their governing implications. Besides the critical need for a national transportation strategy designed to assist municipalities to invest in more sustainable infrastructure, an issue-driven federal government model might more easily be able to implement both sectoral and cross-sectoral strategies such as a national water strategy, an energy security strategy, a climate change adaptation, and mitigation strategy, a waste management strategy, and a climate change, and health security strategy to name only a few. Clearly, a "national" instead of merely a "federal" governance system must have the ability to consider issues outside their institution, in a dynamically interconnected system of governance with the private sector and civil society.

What is crucially required is federal alignment between political accountability that is, the Speech from the Throne, with priority issues, cabinet committee structures, policy development, legislation (change wording of departmental acts), and supportive resource allocations. Such a model is depicted below, an expanded system of governance based on feedback loops from trans-disciplinary networks of collaboration to determine emerging policy domains to identify the domains of appreciation to determine strategic and priority issues of concern. For example, priority issues have just been identified by 45 policy experts who spent two years brainstorming for ways to save the planet,[39] which the Government would be able to use to set the

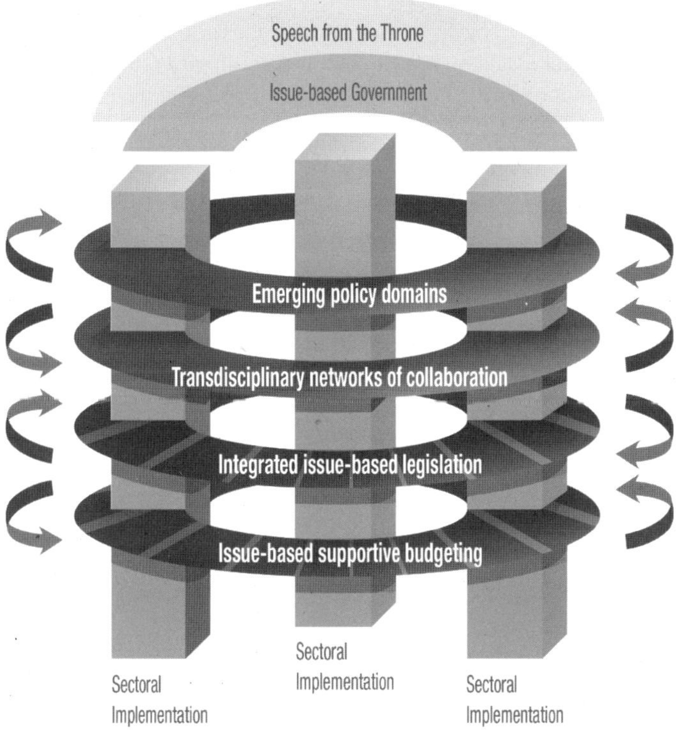

Figure 1
A Networked Governance Model

domains of appreciation in the Speech from the Throne, or if the new model for governance was adopted, its policy experts would have identified this as an important source of information in its policy development process, through their trans-disciplinary networks of collaboration.

I would argue that sustainable development is the sole priority domain, given the definition of sustainable development as a process of reconciliation of three imperatives – the ecological imperative to live within global biophysical carrying capacity and to maintain biodiversity; the social imperative to ensure the development of democratic systems of governance that can effectively propagate and sustain the values by which people wish to live, and the economic imperative to ensure that basic needs are met worldwide. And equitable access to these resources – ecological, social, and economic – is fundamental to its implementation. I would further argue that given the recent evidence released by the Intergovernmental Panel on Climate Change that in addition to being a broad enough imperative now to encompass all government operations, it is primordial.

For purposes of illustration, assume the Government identified priority issues such as climate change adaptation and mitigation, energy security, water security, and sustainable community development. Cabinet committees would then align around these issues, supported by a corresponding Privy Council secretariat. One direct result of the cabinet committee structure being dynamically connected to prioritized and clearly accountable issues is that the vested interests lobbying to maintain the status quo lose a continuity of constituency. Moreover, it is a first step to more adaptive management strategies for departmental organization. Government departments would then be mandated to identify horizontal cross-departmental policy circles to develop policies and to design policy implementation strategies and vertical operational requirements to support their implementation. Given the characteristics of the domain as articulated earlier in this chapter, integrated horizontal policy-making would be needed. For example, sustainable community development, inclusive of the social imperative, would encompass poverty and homelessness besides economic diversification, as well as food safety and access. It therefore involves coherent policy development between Health Canada, Environment Canada, Justice Canada, and Agriculture Canada at a minimum. The Privy Council Office would be responsible for coordinating internal policy circles that cut across sectoral departments, although senior policy analysts would still report to their home department, but policy would be developed using a matrix management model.

In effect, a passive dismantling of stovepipes would evolve, especially with embedded mandates of policy congruence and policy alignment. These policy circles would be complimented by networks of trans-disciplinary practitioners and experts, other quasi-government organizations, and civil society organizations to provide iterative feedback on developing policies and as well, to build in a strategic policy analytical capacity for emerging societal issues for the political level. Ideally they would also create novel constituencies of support for concrete implementation of government policy and the identification to potential and actual barriers to action. These networks would also provide direct evaluation of the success or failure of government policy, feeding back into the system critical information about current and anticipated policy development and identifying emerging policy domains for input into preparations for the Speech from the Throne.

All cabinet memoranda would consequently be assessed against the following checklist – the "fit" with the domain(s) established by the Speech from the Throne; evidence of integrated decision-making; policy and legal alignment across vertical operations; and policy congruence with other levels of governments. Real reform will not occur without corresponding supportive legislation and budgeting, again issue-based, with funding mechanisms. For example, it is now very clear among researchers and some business leaders, and I conjecture the Canadian public, that we cannot fund the necessary sustainable transportation systems and infrastructure, without the immediate implementation of a carbon tax. There is also an equity issue involved for Canadian communities that can only be addressed through a national taxation scheme, as those communities who have heavily invested in less sustainable infrastructure, need transition strategies and financing options to move to more sustainable infrastructure choices.[40]

Let us assume the Speech from the Throne committed to a national water security strategy. How would an expanded decision-making and open policy development process operate? This section draws heavily on a paper written by Bradford,[41] which examined innovative practices and enabling policies in 11 cities in Canada, Europe, and the United States. Assuming that "place-sensitive" modes of policy implementation and programming are necessary and sufficient, and evidence-based policy-making is critical, then the first step in such a process would be to conduct a socio-ecological-political map of the issue at hand, focusing on place specific case studies. Such a mapping would include an environmental scan of the state-of-the art current interdisciplinary research on the sustainable development aspects of the issue – the ecological, social, and economic conditions. Who are the key national stakeholders, government actors, researchers, and networks of organizations involved in the public policy issue? Such a mapping should ideally reveal key individuals and leaders (entrepreneurial nodes) that bridge existing networks, as well as possible institutional intermediaries, from both national, provincial and municipal communities. The political map would identify the Federal government departments and agencies with responsibility for water, any provincial and municipal authorities involved, as well a mapping of the policy, legislative, and regulatory frameworks in place.

The next step would be to convene a policy round table of representative individuals identified from the socio-ecological mapping with all entrepreneurial nodes participating, if possible. Agreement on the prepared environmental scan is then necessary. Next, the goals and priorities of the round table would be agreed upon, with one objective being a social learning process[42] by the members themselves, based on the perspective of the power to reach consensus through deliberation.[43] In this example, the goal would be to develop the vision, goals and objectives of a water strategy, as well as basic principles. One such principle might be territorial equity.[44] Responsibility for developing the final strategy would rest with government; however, all deliberations of the round table would be publicly accessible.

After deciding upon the fundamentals of what should be in the strategy, the round table would then decide on perhaps one of the most essential outcomes of an expanded decision making context, the development of a detailed implementation strategy. Part of this next step would be to determine the institutional infrastructure that should be in place to support the implementation of the water strategy, but respecting its principles. It would recommend what key private-public partnerships, civic alliances, and network clustering would be needed to support, for example, the principle of equitable implementation across communities. In addition, it would identify policy alignments that must take place to realize the strategy, and necessary congruence between levels of government. Another purpose would be to identify gaps in institutional infrastructure that would need to be addressed, and institutional reform with respective accountabilities and resource allocation.

CONCLUSIONS

The basic distinction between governing in the 19th century and today is that our future will depend on government institutions that, aside from mediating the relationship between society and the economy, simultaneously ensure the continuation of ecosystem functional processes.[45] We also have to move beyond mere continuation of these processes to active regeneration of place in many communities. Previous resource management systems based on maximal, instead of optimal scales are unsustainable; institutions must go far beyond current policy skills and knowledge to begin to identify matching, and often nesting, of ecological and social systems and applicable scales of space and time.[46] The necessary skills and knowledge will be available if public servants are supported by a flexible machinery of government that is adaptive and that has in place abridged feedback loops between learning and amending institutional machinery.[47] None of this will happen, however, unless there is renewed political will and leadership based on unprecedented ways of cooperating, to restore the space within the public service for strategic public policy capacity, and renewed trust between the political and public levels.

I am indebted to conversations with my colleagues, Dr. Jim MacNeill, Jim Hamilton and Robert Hilton.

NOTES

1 World Commission on Environment and Development (WCED), *Our Common Future* (Oxford University Press, 1987), 310.

2 A. Dale, *At the Edge: Sustainable Development in the 21st Century* (University of British Columbia Press, 2001); C. Sabel, "A Quiet Revolution of Democratic governance: Towards Democratic Experimentalism," *Governance in the 21st Century* (OECD, 2001); O.R. Young and K. von Moltke, "Governance without government," *Work in Progress of the United Nations University* 14.2 (December 1993); R. Kemp, S. Parto, and R.B. Gibson, "Governance for sustainable development: moving from theory to practice," *International Journal of Sustainable Development* 8.1/2 (2005), 12–30.

3 N. Bradford, *Cities and Communities that Work: Innovative Practices, Enabling Policies,* (Canadian Policy Research Networks, 2003); A. Dale, *At the Edge* (2001).

4 E. Slack, *Municipal Finance and the Pattern of Urban Growth*, C.D. Howe Commentary 160 (2002), 1–25.

5 Young and von Moltke (1993) discriminate between governance as centering on the management of complex interdependencies among many different actors – individuals, corporations, interest groups, nation states – involved in interactive decision-making that affect each other's welfare. Governments, by contrast, are organizations – material entities possessing offices, personnel, equipment and budgets and are often infused with powerful political ideologies.

6 Commission on Global Governance, *Our Global Neighbourhood* (Oxford University Press, 1995).

7 E-dialogues for Sustainable Development. Available at www.e-dialogues.ca.

8 Dale, *At the Edge* (2001); L.H. Gunderson and C.S. Holling, *Panarchy: Understanding Transformations in Human and Natural Systems* (Island Press, 2002); R.B. Norgaard, *Development Betrayed: The End of Progress and a Co-Evolutionary Revisioning of the Future* (Routledge Press, 1994); "Ecosystems and Human Well-Being," in R. Hassin, R. Scholes, and N. Ash, eds. *Millennium Ecosystem Assessment* (Island Press, 2005); J.B. Robinson and J. Tinker, "Reconciling ecological, economic and social imperatives: a new conceptual framework," in T. Schrecker, ed. *Surviving Globalism: Social and Environmental Dimensions* (Macmillan, 1997); Worldwatch Institute, *Vital Signs 2007–2008* (W.W. Norton & Company, 2007).

9 Dale, *At the Edge* (2001).

10 A. Dale and J.B. Robinson, *Achieving Sustainable Development* (University of British Columbia Press, 1995).

11 Jim MacNeill, personal communication, 14 April 2007.

12 J. B. Robinson, G. Francis, R. Legge, and S. Lerner, "Defining a sustainable society: values, principles and definitions," *Alternatives* 17 (1990), 36–46.

13 S.O. Funtowicz, and J. Ravetz, "Science for the post-normal age," Futures 25.7 (1993), 735–55; S.O. Fuctowicz and J. Ravetz, "A New Scientific Methodology for Global Environmental Issues," in R. Costanza, ed. *Ecological Economics: The Science and Management of Sustainability* (Columbia University Press, 1991), 137–52.

14 C.S. Holing, *An Ecologist's View of the Malthusian Conflict* (The Royal Swedish Academy of Sciences Stockholm, 1993)

15 J.S. Dryzek, "Green reason: communicative ethics for the biosphere," *Environmental Ethics* 12.3 (1990), 195–210.

16 Dale, *At the Edge* (2001).

17 Whistler Forum, *A National Agenda for Sustainable Communities*, 19–20 September 2007.

18 C. Sabel, "A Quiet Revolution of Democratic Governance: Towards Democratic Experimentalism," *Governance in the 21st Century* (OECD, 2001).

19 A.N. Whitehead, *Process and Reality* (Harper Brothers, 1929).

20 J.T. Klein, "Interdisciplinarity and complexity: An evolving relationship" ECO 6.1–2 (2004), 4.

21 F. Berkes and C. Folke, eds. *Linking Social and Ecological Systems: Management Practices and Social Mechanisms for Building Resilience.* (Cambridge University Press,

1998); D. Waltner-Towes and J. Kay, "The evolution of an ecosystem approach: the diamond schematic and an adaptive methodology for ecosystem sustainability and health," *Ecology and Society* 10.1 (2005), 3. Available at: www.ecologyandsociety.org/vol10/iss1/art38/.

22 C.S. Holling, *Adaptive Environmental Assessment and Management* (Wiley, 1978); C.J. Walters, *Adaptive Management of Renewable Resources* (McGraw Hill, 1986).

23 E.L. Trist, "Referent organizations and the development of interorganizational domains," *Human Relations* 36 (1983), 269–84.

24 A. Dale and T. Naylor, "Dialogue and public space: An exploration of radio and information communications technologies," *Canadian Journal of Political Science* 3.1 (2005), 203–25.

25 Dale, *At the Edge* (2001); Robinson and Tinker, "Reconciling ecological, economic and social imperatives" (1997).

26 R. Kemp, S. Parto, and R.B. Gibson, "Governance for sustainable development" (2005).

27 Ibid.

28 J. MacNeill, "The forgotten imperative of sustainable development," speech delivered at Pace University, April 20, 2006.

29 Dale, *At the Edge* (2001).

30 P. Hawken, *The Ecology of Commerce: How Business Can Save the Planet* (Weidenfeld & Nicholson, 1993); R. Costanza, R. d'Arge, and R. de Groot, "The value of the world's ecosystem services and natural capital," *Nature* 387 (1997); G.G. Daly, *Nature's Services: Societal Dependence on Natural Ecosystems* (Island Press, 1997).

31 Dale and Robinson, *Achieving Sustainable Development* (1995).

32 J-M. Fontan, "Partnership-oriented research on the social economy in Quebec," *Horizons* 8.2 (2006), 17–19.

33 McPfee and Weibe (1986), cited in K.S. Hanna and D.S. Slocombe, eds. *Integrated Resource and Environmental Management: Concepts and Practice* (Oxford University Press, 2007).

34 A. Dale and J. Hamilton, *Sustainable Infrastructure: Implications for Canada's Future Infrastructure* (March 2007). Available at: www.crcresearch.org.

35 OECD, *Governance for Sustainable Development: Five OECD Case Studies* (2002), 11.

36 Dale and Hamilton, *Sustainable Infrastructure* (2007).

37 Bradford, *Cities and Communities that Work* (2003), 2.

38 M. McMahon and R. Oddie, "Urban Sustainability and Environmental Research in Canada: Prospects for Overcoming Disciplinary Divides?" Commissioned by Oliver Coutard, Directeur adjoint, Laboratoire Techniques, Territoires et Societes, for the Centre national de la Recherche Scientifique, France (York University, City Institute, 2007).

39 Institute for Research on Public Policy. Available at: www.irpp.org.

40 Dale and Hamilton, *Sustainable Infrastructure* (2007).

41 Bradford, *Cities and Communities that Work* (2003).

42 Gordon Smith points out the necessity to shorten the feedback loop between learning and amending institutional machinery. G. Smith, *New Challenges for High Level*

Leadership Training in Public Management and Governance in a Globalizing World (unpublished paper, 2002).

43 M. Piore, *Beyond Individualism* (Harvard University Press, 1995).

44 Bradford, *Cities and Communities that Work* (2003).

45 D.J. Brunckhorst, "Institutions to sustain ecological and social systems," *Ecological Management & Restoration* 3.2 (August 2002).

46 M. McKean, "Common Property Regimes: Moving from Inside to Outside," in B.J. McCay, and B. Jones, eds. *Proceedings of the Workshop on Future Directions for Common Property Theory and Research* (Rutgers University, 1997); M. McKean, "Common-Property Regimes as a Solution to Problems of Scale and Linkage," in S. Hanna, C. Folke, and K.G. Maler, eds. *Rights to Nature: Ecological, Economic, Cultural and Political Principles of Institutions for the Environment* (Island Press, 1996).

47 G. Smith, *New Challenges for High Level Leadership Training in Public Management and Governance in a Globalizing World* (unpublished paper, 2002).

5 Polls, Politics, and Sustainability

MARK S. WINFIELD

The transformation of industrialized, resource dependent and urbanized societies like Canada's in the direction of environmental sustainability presents
significant challenges. Placing our economy and society on a more sustainable basis will require major changes in current patterns of energy and materials production and use, while maintaining economic prosperity and social
well-being. It has been estimated, for example, that to achieve sustainability
worldwide, the material intensity of each unit of economic output will need
to be reduced by 50 percent and, in industrial countries like Canada, it will
have to fall by factors of between 4 and 10.[1] Reductions in emissions of
greenhouse gases on the order of 80 percent relative to 1990 levels will be required by the middle of this century.[2]

Changes in patterns of materials and energy production and use on this
scale imply significant changes in the current structure of the economy. The
achievement of such changes would require substantial changes in existing
public policies and institutional arrangements around the management of
the environment and natural resources. In the Canadian case, the most striking feature of our responses since the emergence of the environment as a distinct public policy issue now more than forty years ago has been the lack of
progress in the achievement of these sorts of changes.

There is no doubt that Canada has made substantial progress in a number of
specific areas. Prominent examples include the almost universal implementation of municipal sewage treatment in the Great Lakes basin; acid rain control
in eastern Canada; reductions in water pollution from the pulp and paper sector; and the phase-out of the production of certain harmful substances, such

as ozone depleting substances (ODS) and polychlorinated biphenols (PCBs). Unfortunately, however, Canada's overall trajectory in terms of production and consumption of energy and materials remains fundamentally unaltered, at least in terms of the results of any environmental policy interventions over the past forty years.[3]

Despite fifteen years of international commitments and domestic policy initiatives to stabilize or reduce greenhouse gas emissions, for example, Canada's GHG emissions have risen steadily. Indeed, Canada's 2004 emissions were 34.6 percent higher than its target for the 2008–12 period under the 1997 Kyoto Protocol to the 1992 United Nations Framework Convention on Climate Change (UNFCC).[4] The federal office of the Commissioner for the Environment and Sustainable Development (CESD) has repeatedly highlighted the overall failure of Canadian governments to implement international and domestic environmental policy commitments.[5] The Organization for Economic Cooperation and Development (OECD), for its part, has criticized Canada's continued subsidization of non-renewable resource exploration and extraction and failure to make significant use of economic instruments to move consumption patterns of water or other resources in more sustainable directions, on a number of occasions.[6] Instead, Canadian environmental policy has been dominated by issue specific initiatives, often with relatively short time horizons and limited enduring impact. Such approaches are unlikely to alter the fundamental trajectory of Canada's patterns of materials and energy production and use in the direction of greater sustainability.

SUSTAINABILTY AND POLITICAL TIME FRAMES

It is widely accepted that the transition towards sustainability will require changes in policies and institutions that will take a long time to translate into visible improvements in environmental quality. In this context, the lack of progress is often explained in terms of the relatively short timeframes within which decision-making occurs at the political level in a Canadian context. Decision-making horizons are dominated by the four to five-year electoral cycle that defines the lives of Canadian governments, and the even shorter time periods over which individual ministers typically hold their portfolios. In the case of environment ministers, tenure with the portfolio is typically two years or less.

In these circumstances, decision-makers may feel little incentive to invest in policy measures whose benefits may not be visible until long after they have left their portfolios and the governments of which they are part have left office. Even where environment ministers may be convinced that policy investments in the long-term make sense, they find it a very hard sell to their cabinet colleagues, particularly where there is the potential for significant transitional costs affecting specific industries or constituencies. Rather, the system provides incentives to opt for initiatives that will yield short-term political benefits, but may do little to advance sustainability in the long term. The Government of Ontario's recent proposal to make "green" licence plates available for hydrid and other high efficiency vehicles,[7] while failing to include more substantial initiatives to improve the fuel economy of the province's vehicles through aggressive fiscal incentives, to say nothing of regulatory requirements, in its climate change strategy is a recent example of such behaviour.

A diagnosis of environmental policy failure as a result of a short electoral cycle-driven policy attention deficit disorder on the part of politicians suggests solutions that must turn on convincing decision-makers to act on the basis of much longer time frames, even though such behaviour may not translate into electoral benefits to them or their governments. Such an approach presents a tall political order, as it may involve asking politicians to ignore their basic instincts for political survival.

At the same time, the diagnosis may underestimate the complexity of the political and policy challenges presented by environmental sustainability. In addition to its vulnerability to short electoral cycles and even shorter ministerial tenures, it is also apparent that there is a strong relationship between the public salience of environmental issues and the level of governmental activity in relation to them. Kathryn Harrison's classic illustration of the correlation between levels of public concern for the environment and Environment Canada's budget from the mid-1980s to the early 1990s below demonstrates this point clearly. [8]

Government spending and policy activity on the environment increase notably during periods of high public concern. During such periods of high concern for the environment, politicians perceive a potential for political rewards for action, and also political risk if they are seen as failing to respond.

In contrast, it can be extraordinarily difficult to move significant policy change forward during periods of lower public salience. The resistive capacity of interests who may be potentially adversely affected in the short term by environmental policy initiatives is usually simply too overwhelming to be overcome. Affected interests typically enjoy advantages of resources, institutional support from government agencies and ministers for who they are important clientele, and ultimately the "structural' power of economic interests to threaten to withdrawn investment, and by implication employment from specific communities. Even very capable ministers of the environment may find it virtually impossible to prevail over these barriers at the cabinet table without ability to argue the possibility of significant political advantages arising from taking action or the potential for major electoral costs if effective action is not seen to be taken.[9] This is not an iron rule. Upon occasion, initiatives may slip through when their significance is not recognized at the time.

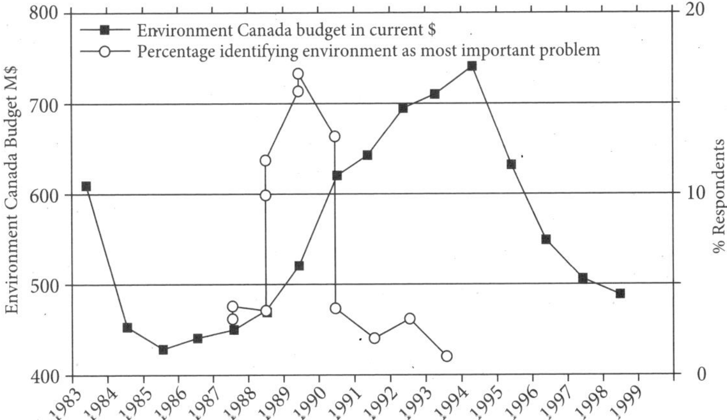

Figure 1

The correlation between levels of public concern for the environment and Environment Canada's budget from the mid-1980s to the early 1990s.

The adoption of the Federal Environmental Assessment Guidelines Order in 1984 would be an example of such an outcome. However, in general, unless some other external political factors are at play, such as minority government situations where one or more of the opposition parties takes a strong interest in environmental issues (Ontario 1975–81),[10] or prime ministerial interventions based on international pressure or "legacy" concerns (the Kyoto negotiating position (1997) and ratification (2002) decisions)[11] substantial policy change is extremely difficult to move forward outside of periods of high public concern for the issue.

The distinct ways in which levels of public concern for the environment shift makes this pattern of dependency on the windows of opportunity to advance major initiatives presented by periods of high public concern particularly problematic. Public concern for the environment in Canada, and indeed, internationally has been characterized by long periods of relatively low salience, interrupted by brief periods of extremely intense public concern.

It is generally accepted that in Canada, and much of the rest of the western world, there have been three of the periods of high public concern for the environment as a public policy issue over the past century. There is also strong evidence that we are in the midst of a fourth such episode. These periods have been the key moments in the establishment of institutions, legislation and policies related to natural resources and the environment at the provincial, national and international levels.

The first identifiable wave of interest emerged around the beginning of the previous century, where concerns over the wasteful and unsustainable consumption of renewable resources, particularly forests, lead the Laurier government to establish a Commission on Conservation (1908–21).[12] The period also saw the emergence of the first international environmental agreements, notably the Canada-us Boundary Waters Treaty of 1909 and the Canada-us Migratory Birds Convention of 1911. In addition, the period witnessed the establishment, by the 1920s, of initial policy and institutional frameworks around the sustainable management, as opposed to the simple maximization of revenues arising from exploitation, of renewable natural resources, particularly forests, at the federal and provincial levels in Canada.[13] The original North American "conservation" movement was pre-eminently grounded in scientific and professional communities, as opposed to being a mass social movement in the sense of modern environmentalism. At the same time, it did have a strong populist element highlighted, for example, by the appeal of us President Theodore Roosevelt's initiatives to establish the us national parks system and us Forest Service.[14]

The second wave of concern occurred during the late 1960s and early 1970s, marking the emergence of the "environment," in the modern sense, as a distinct public policy issue. The second wave is also generally recognized as the time of the birth of modern environmental movement, shifting from the earlier "conservation" movement's greater salience among certain academic and professional elites, to a movement with broad public appeal. During this period much of the current federal and provincial institutional and legislative framework for environmental management in Canada was established, including the formation of ministries and departments of the environment, and the adoption of comprehensive legislation for the regulation of air and water pollution and waste management. Many of Canada's leading environmental non-governmental organizations, such as Pollution Probe (1967), Canadian Environmental Law Association (1970), and West Coast Environmental Law Association (1974) were founded at the same time. At the international level, the 1972 Stockholm Conference on Man and His Environment led to the establishment of United Nations Environment Program and an initial round of global international environmental agreements, such as the International Convention for the Prevention of Pollution from Ships (MARPOL). In North America, the Canada-us Great Lakes Water Quality Agreement of the same year laid the groundwork for an ambitious bi-national agenda for the restoration and protection of the Great Lakes basin.

The third wave of public concern for the environment emerged in the late 1980s and early 1990s, culminating in the World Conference on Environment and Development in 1992. In Canada, the period was marked by the adoption of new environmental legislation at the federal and provincial levels, including the *Canadian Environmental Protection Act, 1988* and the *Canadian Environmental Assessment Act, 1992*. Significant regulatory initiatives were launched at the federal and provincial levels, particularly in relation to acid rain control in eastern Canada and water pollution from the pulp and paper sector. New institutions, including environmental commissioner's offices at the federal level[15] and in Ontario[16] were also created. On the international stage, Canada participated in the development of a host of new international and regional environmental agreements dealing with ozone depleting substances, climate change, biodiversity, transboundary movement of hazardous wastes, air quality and a range of other issues. The period was also notable for explicit efforts, flowing from the work of the World Commission on Environment and Development (The Brundtland Commission)[17] to re-conceptualize the environment-economy relationship. In Canada these efforts were manifested institutionally through the creation of provincial and national round tables on the environment and economy and an International Institute for Sustainable Development.[18]

A fourth wave of public concern began to emerge in the fall 2006, with the highest ever levels of public concern for the environment in Canadian public opinion polling being recorded in January 2007 (35%).[19] The most notable feature of this peak has been the degree to which public concern, at least at the national level, as been very strongly focused on the issue of climate change.[20] Previous peaks have been characterized by a focus on a wider range of specific environmental issues.

Peaks in public concern for the environment have the character of being self-reinforcing events. As public concern grows, the level of media attention given to the issue increases further enhancing public interest, which in turn prompts additional coverage. Unfortunately for the environment the reverse dynamic is also possible, with falling public concern, media interest, and governmental activity combining to produce cycles of declining attention.

Between these periods of high public saliency, the environment has tended to rank relatively poorly among identified top of mind public concerns, typically with less than 5 percent of survey respondents identifying it as a leading issue. Figure 2, for example, highlights the low status of the environment as a top of mind issue in Canada between 1992 and current peak in public concern.

Public opinion survey data from a number of sources does suggest that the environment has become "embedded" as a value in Canadian society, with the result that even during periods of low saliency, when prompted, members of the public will indicate concern and expectations of high levels of governmental action. Surveys conducted in Ontario in 1996, for example, when the status of the environment as a top of mind issue was extremely low, being identified by only 1.2 percent of respondents in a survey taken in the fourth quarter of that year,[21] revealed that more than 80 percent of respondents, when asked, indicated that governments should make environmental laws stricter, even in the context of major reductions in overall government budgets.[22] However, absent evidence of status as top-of-mind concern, this "embedded" concern for the environment and expectation of government action does not seem to translate into a political factor of great consequence. In fact,

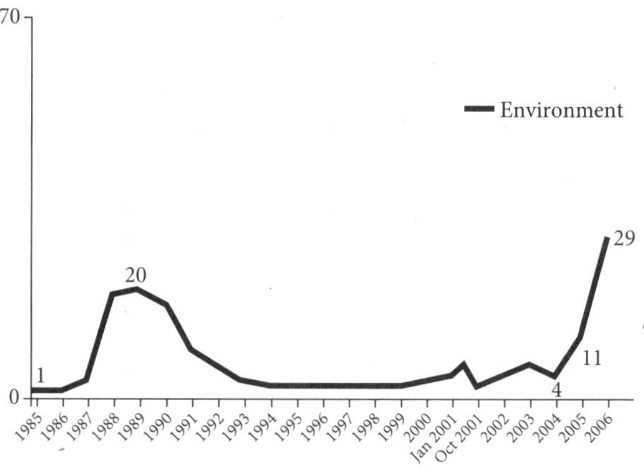

Figure 2
The Environment as Top-of-Mind Issue 1985–2006[76]

in Ontario, the provincial government began to implement a reduction of more than 40 percent in the Ministry of the Environment's operating budget and a major regulatory "reform" initiative whose primary feature was to weaken regulatory requirements in relation to environmental protection, within a few months of the completion of that survey.[23]

The pattern of brief periods of intense public interest, followed by long periods of inattention has been sufficiently pronounced both in Canada and internationally with respect to the environment that the issue has been used to illustrate the concept of an "issue attention cycle."[24] The term is used to describe cycles of government activism around an issue coinciding with high levels of public concern or interest, and a falling off of government action as public interest in an issue declines. Although similar patterns of behaviour have been witnessed with other policy issues,[25] there may well be an element of exceptionalism around the degree to which environmental policy activity is sensitive to the public salience of the issue. The environmental policy case may reflect a number of factors, including the capacity of interests committed to conventional economic development patterns to reestablish existing policy directions absent strong political motivations for policy action. The passing nature of the policy responses from governments, which have tended to avoid long-term institutional changes and policy commitments which might have the effect of reinforcing the level of public attention given to environmental issues, have been another factor.

The cyclical nature of public interest in the environment presents significant problems in terms of the formulation of public policies with lasting impact. With each wave of public concern, governments have become increasingly conscious of their temporary nature. In some cases this may prompt efforts, particularly on the part of environment ministers, to make the most of the opportunities presented during waves of concern. However, it can also lead to policy responses that are intended to satisfy public concern by providing the appearance of action, while avoiding any fundamental shift in the underlying direction of policies and institutions towards economic development as conventionally understood.

From the perspective of political decision-makers, this can be perceived as a rational strategy in terms of both policy and politics. If one views the level of public concern and interest in the environment as being fundamentally transitory in nature, with the result that political rewards for action may not be there to be collected by election time, then there are strong incentives not to undertake more fundamental policy changes which may open the possibility of other political risks to the government. Such risks may include short-term job losses or disruptions of the activities of important constituencies (e.g. resource related employment in one industry towns). Governments may be particularly inclined to pursue such strategies if their underlying perception of the seriousness of environmental problems is weak to begin with.

The result is typically temporary or one time only spending and program initiatives that end when their funding allocations expire (if the announced funds are ever spent at all).[26] The 1990 federal Green Plan provides a prominent example of this type of policy response, where $3 billion was committed, but little lasting legacy remained, particularly outside of Environment Canada, once what portion of those funds remained after several rounds of budget cuts, was spent.[27] The fundamental direction of departmental or ministry A base budgets, and the basic fiscal and taxation structure that continued to encourage and facilitate conventional economic models remained untouched. A recent review of provincial GHG reduction strategies revealed a similar pattern. Aside from the introduction of a carbon tax in British Columbia,[28] and in a more modest form in Quebec, the strategies are dominated by one-time only spending initiatives.[29]

A prominent example of the failure to affect changes to underlying economic development policies is the continued subsidization by the federal government of extractive non-renewable resource sectors through tax expenditures and other mechanisms. The situation with respect to fossil fuels, the key source of greenhouse gas emissions, is particularly egregious. Despite commitments to the stabilization of GHG emissions through the 1990 Framework Convention on Climate Change, and to a 6 percent reduction in emissions relative to a 1990 baseline under the 1997 Kyoto Protocol, and a succession of national and federal climate change strategies and plans, federal support to the sector, conservatively estimated in 2005 at more than $1.4 billion per year,[30] continued. A phase-out of the accelerated capital cost allowance (ACCA) for oil sands developments, introduced in the 1996 budget, was announced in the 2007 budget.[31] However, ACCA will not be phased out 2015, with the result that it will have little effect on existing and planned oil sands developments, key drivers of Canada's increases in GHG emission levels.

In some cases, where a particular issue has very high salience, such as acid rain or water pollution from the pulp and paper sector, it may be the target of regulatory action which has some lasting effect in terms of improvements in environmental quality. However, on the whole, governmental responses, paralleling the governmental perceptions of public concern for the issue, seem increasingly shallow and short lived. So far the current peak in public concern for the environment has not produced the levels of regulatory activity, institutional innovation, and attempts to engage intellectually with environment-economy relationship that characterized the previous three cycles of attention. Recent federal proposals to establish regulatory frameworks related to greenhouse gas emissions and air quality, although offering the prospect of more substantial action, remain nothing more than proposals.[32] British Columbia's recent introduction of North America's first substantial carbon tax is an important exception in this regard.[33] It remains to be seen if other Canadian governments will follow through with initiatives of comparable significance.

The transitory nature of public concern and increasingly transitory governmental responses present serious enough challenges to advancing environmental sustainability. Yet this is not the complete picture. It is also important to note changes in governmental behaviour during the long troughs in public concern for the environment between the peaks. Previous troughs, particularly the one between the mid-1970s and mid-1980s were largely marked by periods of neglect of environmental policy, with occasional episodes of modest innovation and retrenchment.[34]

The most recent trough, however, which reached its nadir in the mid-to late 1990s, coinciding with neo-liberal political and policy "revolutions" at the provincial level, led by Ontario and Alberta, and the program review experience at the federal level, was not limited to a halt in progressive action, or even simple neglect. Rather this period witnessed the actual dismantling of some of the legislative and institutional structures built up over the preceding waves of public concern, including provincial round tables on the environment and economy and reductions of the operating budgets of environment ministries and departments of between 30 percent (Environment Canada and Alberta) and nearly 70 percent (Quebec, British Columbia, and Newfoundland).[35] The severity of this most recent trough may have been an anomaly – the product of a near "perfect storm" against environmental policy activism of very low public salience, fiscal "crises" at the federal and provincial levels, a crisis of federalism prompted by the results of the 1995 Quebec referendum, and the dominance of neo-liberal ideas about the role of the state, emphasizing the minimization of governmental activity and interventions in the economy.[36] Indeed, the only comparable episode of retrenchment in terms of governmental capacity in the environmental and natural resources management field occurred during the depression of the 1930s.[37] The events of the late 1990s highlighted the possibility of severe retrenchment rather than a simple stalling of progress during periods of low public salience of the issue.

The challenges presented by these realities of the interaction of public opinion and governmental behaviour with respect to environmental sustainability are substantial. They imply that there may only be relatively narrow windows of opportunity during which major initiatives can be advanced. More hopefully, the issues of the environment and sustainability do have certain important features that may provide the foundation for the formulation and implementation of effective long-term policy interventions.

Perhaps the most important of these features is the fundamental empirical grounding of the issues. Concern for the environment has its essential origins in both the experienced effects and scientifically established diagnoses of threats to human health, safety and well-being and the structure and health of the local and regional ecosystems and the global biosphere. The scientific community continues to generate bodies of evidence of problems even absent strong governmental interest. The initial stages of both the global climate change and ozone depletion issues are among the best examples of this phenomenon. In addition, at times, such as during the drinking water contamination episodes in Walkerton, Ontario, and North Battleford, Saskatchewan, Mountain Pine Beetle infestations in British Columbia and Alberta, extreme weather events, and smog episodes in southern Ontario and New Brunswick, environmental problems become too obvious, and the political cost of inaction becomes too great, for governments to be seen to be ignoring them.

Secondly, notwithstanding the issue attention cycle character of the status of the environment as a top of mind policy concern among the public, longterm public opinion survey research supports the hypothesis that the environment has become an increasingly embedded "core" value in Canadian society. Strong latent concern for the environment remains in place among the public even during periods of very low top-of-mind salience. The deepening of the "embeddedness" of the environment as a public value may be reflected in the fact that the levels of top of mind concern have become higher with each of the three modern peaks in concern. Polling data from the early 1970s suggests a peak of approximately 6 percent of respondents identifying "pollution" as the most important problem facing Canada,[38] compared to 21 percent of respondents identifying the "environment" as the leading concern in 1989 and 29 percent of respondents identifying the "environment" as the leading concern in 2006.[39]

A third factor, flowing from the second, is a very strong underlying expectation of governmental action with respect to the protection of public goods in general, and the environment in particular. Notwithstanding the direction of government policies, throughout the 1990s there was strong and consistent public opinion polling evidence indicating that the public never accepted a minimalist role for government in the protection of public goods. Rather the public continued to expect governments to protect the environment, and public health and safety, and to employ compulsive (i.e. regulatory) instruments in doing so.[40] The governments involved were themselves always careful to deny that they were engaged in activities that would weaken the protection of public goods.

IMPLICATIONS FOR POLICY AND STRATEGY

The three elements of an empirical grounding in science and experience, increasingly deeply embedded public concern, and ongoing expectations of governmental action, can provide the political foundation of future environmental policies. In this context a key goal of future strategies must be to alter the current pattern of environmental policy development of long periods of relative inactivity or even retrenchment, punctuated by brief periods of intense policy activity corresponding with peaks in public concern for the issue. Canada is more likely to make significant progress on improving the environmental sustainability of its economy and society if environmental policy is developed less in fits and starts, with occasional bouts of reversal, and more on a basis of steady progress. The pattern needs to be one of ongoing evolution, supported by strongly embedded levels of public concern, rather than the "punctuated equilibrium"[41] of the past four decades.

In terms of public opinion, there is evidence that the pattern of intense peaks and long troughs in public concern can be attenuated. In the case of Ontario, for example, the environment has consistently occupied a position among the top five overall issues of public concern since the late 1990s.[42] Since then the issue has regularly ranked third among public concerns in Ontario, after health care and education.[43] As of July 2007 it was tied with heath care as the leading vote determining issue.[44] A number of factors contributed to this more prominent status for the issue: the public interventions of health professions, particularly the Ontario Medical Association[45] and medical officers of health[46] around the health impacts of air quality issues in southern Ontario from 1998 onwards; the May 2000 Walkerton drinking water disaster;[47] and the increasing prevalence of severe smog episodes affecting a large area of southern Ontario each summer.[48]

In political terms this sustained level of attention has translated into the emergence of a substantial body of support, consistently in the range of 7–8%[49] of respondents to polls regarding party preference, for the provincial Green Party from the fall of 2004 onwards.[50] Support at this level carried through in the party's popular vote in the 2007 provincial election,[51] although the party failed to elect any members. The 2007 election outcome suggests the potential emergence of an electorally significant block of voters for whom the environment is a vote determining issue in Ontario. It remains to be seen whether these levels of support will be maintained over the long-term. If support for the Green Party endures, it may have a stabilizing effect on the pace of environmental policy formulation and implementation, given the prominence of the issue it suggests in an electoral context. There are suggestions that the parliamentary presence of green parties in the European Union has had a positive and stabilizing effect on the pace of environmental policy development there.[52]

Unfortunately in a Canadian context, the translation of expressions of preferences in public opinion polls, or even shares of the popular vote into elected members of legislatures still presents major challenges for the Green Party. The existing first past the post electoral system works strongly against parties whose support is relatively widely distributed geographically, as is the case with the green party. Moreover, given the tendency of the existing system to produce majority governments, election of one or even a handful of members will likely be insufficient to significantly affect the direction of legislative development. Rather the election of enough members to obtain party status in a legislature or parliament is required for that. The implication, at least in the short term, is that the most important impact of the portion of the popular vote obtained by the Ontario greens in the October 2007 provincial election may be to make apparent to other parties the presence of a block of electors prepared to vote and to make voting decisions on the basis of environmental issues. Exit polling in Ontario found that forty-five percent of those who voted green in the 2007 election indicated that the environment was the vote determining issue for them. This was by far the highest correlation between a particular issue and party choice among Ontario voters in the election.[53] The outcome suggests the emergence of a core body of green supporters for whom the environment is the determining issue for voting decisions, rather than the party simply being a "parking spot" for voters discontented with the Liberals, Progressive Conservatives and New Democrats.

With respect to the policy impact of the emergence of a consistent and substantial level of support for the Greens in Ontario, meaningful assessments of the environmental policy aspects of the late stages of the Harris and then Eves (2002–03) Progressive Conservative governments, and the first (2003–07) McGuinty Liberal government have yet to be developed. However, the sustained levels of public concern for the environment in Ontario since the late 1990s have coincided with an extended period of relatively high levels of policy activity on the part of the Progressive Conservative and then Liberal governments, particularly with respect to drinking water quality[54] and the management of urban growth in the Toronto region.[55] At the same time, the province's performance in other areas, particularly electricity and energy policy has been the subject to strong and sustained criticism from environmental advocates.[56]

Making the most of major peaks in public concern for the environment, as again seems to be occurring, is critical to laying the groundwork for a long-term shift in the patterns of environmental policy formulation in Canada. The historical record does suggest that these peaks are the key windows of opportunity for policy and institutional innovation with respect to the environment. It is therefore extremely important when there is a preparedness on the part of decision-makers to invest political capital in environmental initiatives, that it be invested in initiatives that will have lasting impacts. Advocates of environmental sustainability and effective environmental policies need to raise the level of expectations around the quality of the interventions offered by governments, and make it clear that symbolic and transitory responses will not yield the expected political rewards or required policy outcomes. Policy advocacy by Canadian ENGOs in recent years has focused strongly on the achievement of micro-policy outcomes related for example, to specific substances, species or natural heritage sites. Greater emphasis needs to be placed on more strategic policy interventions that will affect a wide range of specific issues.[57] I suggest six areas where investments of political capital would be particularly useful.

Pursue Serious Ecological Fiscal Reform

Canada's existing fiscal regime embeds powerful incentives against environmental sustainability. Extensive supports are provided through the federal and provincial tax and resource royalty systems for the development and exploitation of non-renewable resources, particularly fossil fuels and metals.[58] The effect of these incentives is to artificially lower the cost of using these resources in the marketplace relative to more sustainable options, be they increased energy and materials efficiency or low-impact renewable energy resources. The inertia of these long-standing regimes works to counteract the impact of policy interventions intended to promote sustainability.

In this context, significant ecological fiscal reform will be essential to achieving the long term structural changes in economic activity essential to moving environmental sustainability forward. Removing subsidies and internalizing the externalized environmental costs of the provision of goods and services, is not only economically efficient, but will also signal long-term shifts in policy direction to private capital. The transitory nature of the temporary or one-time only initiatives that have characterized most fiscal responses on environmental issues, on the other hand, send the opposite signal – that concern for the environment is short-term, and that the fundamental orientation of public policy towards non-renewable resource development remains unaltered.

Shifts in the basic costs of goods and services to reflect their full environmental costs, through, for example, the introduction of a tax on the carbon content of fuels, will also tend to have systemic impacts on the economy. Such approaches are therefore well suited to achieving the types of structural changes in economic behaviour needed over time to advance sustainability. The political capital available during peak periods of public concern for the environment would be better spent on structural initiatives such as removing perverse subsidies in such areas as fossil fuel and non-renewable resource development, as repeatedly recommended by the OCED and others, than on short term and one-time only expenditures. British Columbia and Quebec's recent introductions of carbon taxes represent important developments in this regard.

Don't Be Afraid to Use Regulatory Tools

Much of the environmental policy literature of the past decade has vilified the use of "command and control" regulatory instruments as cumbersome, outdated and spent.[59] A considerable infrastructure of "regulatory management" has been constructed to effectively prevent their use, particularly at the federal level. [60] The federal government's recent "smart" regulation initiative has reinforced these trends.[61]

In reality, regulation has been the most effective tool employed in the Canadian environmental policy experience and around the world. Much of the measurable improvements in environmental quality that can be attributed to policy interventions over the past forty years, including acid rain control, the elimination of lead in gasoline, reductions in industrial water pollution, and the phase out of the production and new uses of ozone depleting substances (ODS) and PCBs, are the direct result of the use of regulatory instruments. In the case of acid rain, for example, regulations adopted in Ontario and Quebec in the 1980s had, by the mid-1990s, reduced industrial sulphur dioxide emissions in eastern Canada to a level 54 percent lower than those seen in 1980.[62] With respect to the pulp and paper sector, following the introduction of new regulations on discharges introduced at the federal and provincial levels in the early 1990s, by 2000 discharges of chlorinated dioxins and furans were reduced by nearly 99 percent, releases of biological oxygen demand materials fell by 94 percent and releases of total suspended solids decreased by 70 percent relative to their 1987 levels.[63]

Legislation and regulations also have the advantage of being difficult, (although not impossible) to undo without attracting significant ENGO, opposition party and media attention, even during periods of low public salience of environmental issues. As both the Harris government in Ontario and the Campbell government in British Columbia discovered in the 1990s and early years of the current decade, attempts to "make it legal to pollute more" are natural targets for criticism.[64] The consequence is an upwards ratcheting effect in terms of environmental performance requirements.

It is crucially important that regulatory tools be used creatively to affect structural changes in the use of materials and energy, as has been the case with extended producer responsibility requirements in the European Union,[65] rather than prompting simple add on, pollution abatement responses to existing process. It is also important to establish mechanisms for the continuous upgrading of standards and requirements. Many us states, for example, including California, attribute much of the success of their energy conservation strategies to implementation of regular (typically three-year) upgrading cycles to their energy efficiency standards and codes for buildings and energy consuming appliances and equipment.[66] The approach taken at the state level in the us stands in stark contrast to the Canadian model for developing and implementing regulatory standards, be they the energy efficiency requirements of building codes or pollutant emission standards. In Canada, the practice has been to go through intense periods of consultation and lobbying around the development of standards and then leaving the resulting requirement static with no tangible expectation of review or updating. The energy efficiency provisions of Ontario's building code, for example, had been left untouched for more than a decade prior to the revision that occurred in 2006.[67] Canada's federal standards for water pollution from the pulp and paper industry have been left unrevised for more than 15 years.[68]

Embed Commitments to Action in International Agreements.

The incorporation of commitments to action into international agreements can have powerful impacts on domestic policy. The need to fulfill international commitments, once entered into, can be powerful drivers of domestic action. In a Canadian context, it is difficult to imagine the level of public and media attention given to the climate change issue absent the decisions to sign and ratify the Kyoto Protocol. International commitments carry considerable moral weight in domestic policy in part as they are seen as not being able to be abandoned easily without placing the country's broader international reputation at stake. That said, Canada's performance in relation to its Kyoto obligations makes it clear that international commitments do not, in themselves, guarantee the achievement of the required policy outcomes.

The need to fulfill international obligations can be an important consideration in budgetary discussions within governments. At the same time, the fate of the funding commitments needed to fulfil Canada's obligations under the Canada-us Great Lakes Water Quality Agreement during the federal "program review" budgetary reductions in the mid-1990s makes it apparent that international commitments do not assure budgetary security.[69] International commitments can also embed important reporting requirements on performance which cannot be easily avoided. Domestic "initiatives" and announcements aside, Canada ultimately has to report to unfcc on our progress (or lack thereof) in reducing ghg emissions.

International processes carry with them definitive points at which clear decisions have to be placed on the record regarding whether or not to commit to international initiatives. The decision points with respect to the establishment of negotiating positions, and then the signing and ratification of the Kyoto Protocol are prominent examples of such events.[70] There may be significant opportunities to develop a more ambitious international environmental agenda as the dominance of the non-security related international agenda by trade liberalization declines, as evidenced by the stalling of progress on both World Trade Organization and the Free Trade Area of the Americas initiatives. In particular, the strong commitments to domestic GHG emission reductions by all of the leading candidates in the 2008 US Presidential election,[71] along with the November 2007 defeat of the Howard government in Australia, may fundamentally alter the dynamics of future international negotiations on climate change in favour of more rapid and substantial action. More broadly, the establishment of carbon pricing regimes in the United States and European Union may open debates on the restrictions contained in existing international trade agreements on the use of production and process method based trade restrictions. Recent legislative proposals in the United States, for example have included the possibility of the imposition of carbon tariffs on goods imported from countries without carbon pricing regimes.[72]

*Build Institutions to Assess and Report on Progress and Act
as Incubators for New Ideas.*

The Offices of the Commissioner for Environment and Sustainable Development (CESD) and of the Environmental Commissioner of Ontario (ECO), key survivors of the round of institutional formation that accompanied the third wave of public environmental concern of the late 1980s and early 1990s, have played a crucial role in keeping environmental issues on the public and policy agenda during the long trough in public concern of the 1990s. At the same time, the National Round Table on the Environment and the Economy and the International Institute for Sustainable Development, two other "perfect storm" survivors have played important roles as facilitators and incubators of innovative thinking about sustainability.

However, these institutions are increasingly in need of renewal. The contributions of both the federal and Ontario environmental commissioners" offices could be significantly enhanced if they were provided with explicit mandates to assess the effectiveness of environmental policy initiatives in advancing sustainability, rather than focusing on the "management" of policy in the case of the CESD and the processes by which initiatives are developed in the case of the ECO. Given their independence, these agencies would be logical homes for ongoing state of the environment reporting functions.

The need for structural reform with respect to the CESD was highlighted by the sudden departure of Commissioner Johanne Gelinas in January 2007.

Unfortunately, the subsequent "Green Ribbon Panel" review of the office, beyond recommending that the Commissioner be appointed for a fixed seven-year term, effectively proposed a continuation of the office's approach to its existing mandate.[73] Consideration needs to be given to establishing the CESD as a parliamentary officer, separate from the Office of the Auditor General, and reporting directly to the House of Commons. At a minimum, the office should function within the Office of the Auditor General as the 1995 amendments to the Auditor General Act creating the office intended, with the commissioner reporting directly to the auditor general.[74] The amendments" provisions implied a status for the commissioner as a deputy auditor general, rather than as an assistant deputy auditor general as is currently the case. Status as a deputy auditor general would have significant implications for the independence and autonomy of the office within the OAG hierarchy.

Invest in Environmental Science and Information Dissemination

Continued investments in environmental science are fundamental to building the empirical case for action to both governments and the public, and for assessing the effectiveness of the action we do take. Much greater emphasis needs to be placed on making data gathered by governments accessible in useable forms and on knowledge transfer from university-based researchers to those involved in the development, implementation and assessment of policy.

Invest in Environmental Education

Environmental education and awareness is essential to building constituencies for policy interventions both in the present and future. It is also essential to motivating and sustaining behaviour changes at the individual and household level. The widespread participation of households in increasingly ambitious waste diversion activities, transitioning in some communities, like the City of Toronto, from a simple process of taking bags of garbage to the curb every week, to sorting household wastes into six or seven streams,[75] resulting in diversion rates exceeding forty percent,[76] with little direct economic incentive or regulatory enforcement, highlights the potential impacts on behaviour of effective education and awareness initiatives.

CONCLUSIONS

Environmental policy in Canada is highly sensitive to the public salience of environmental issues. Periods of high levels of public concern for the environment are associated with major legislative, institutional and policy initiatives. In contrast, periods of low public salience are associated with low levels of governmental policy activity and even policy retrenchment.

The dynamic of policy cycles around the environment presents significant challenges in terms of advancing and sustaining the types of policy initiatives needed to move existing patterns of materials and energy production and use in Canada in the direction of greater sustainability. At the same time, the empirical grounding of the environment as a public policy issue in science and the personal experience of members of the public, along with deeply rooted public expectations of governmental action on the issue, provide the foundation for future policy initiatives.

Canadian governments should take the opportunity offered by current and future periods of high public salience of the environment as a policy issue to lay the groundwork for long term economic change towards environmental sustainability through ecological fiscal reform and the creative use of regulatory instruments. At the same time, efforts should be made to strengthen the long term profile of environmental issues through the enhanced generation and access to environmental science and information, the embedding of policy commitments in international agreements, and strengthening mechanisms for independent evaluation and public reporting on environmental progress in Canada. The recent emergence of significant electoral support for green parties at the provincial and federal levels, if sustained, may also have a positive and stabilizing influence on the pace of environmental policy development in Canada.

NOTES

1 The need for a 90% reduction in material intensity in OECD countries was acknowledged in the October 1994 Carnoules Declaration, endorsed by prominent individuals including the former executive directors of the Business Council for Sustainable Development and the Brundtland Commission. See T. Green, *Lasting Benefits from Beneath the Earth* (1998), 69. See also G. Gardner and P. Sampat, *Mind over Matter: Recasting the Role of Materials in Our Lives* (Worldwatch Institute, 1998); J.Young and Aaron Sachs, *The Next Efficiency Revolution: Creating a Sustainable Materials Economy* (Worldwatch Institute, 1994); Fresenius Environmental Bulletin (special edition on The Material Intensity Per Unit of Service (MIPS) project of the Wuppertal Institute, Germany) 2.8 (1993).

2 See for example, M.Bramley, *The Case for Deep Reductions: Canada's Role in Preventing Dangerous Climate Change* (Pembina Institute and David Suzuki Foundation, 2005).

3 There has certainly been structural change in Canada's economy over this period, but it has not in any sense been the result of domestic environmental policy interventions.

4 Commissioner for Environment and Sustainable Development (CESD), *Report of the Commissioner for the Environment and Sustainable Development* (2006) Available at: www.oag-bvg.gc.ca/domino/reports.nsf/html/c20060900se01.html; accessed September 14, 2007.

5 See CESD, "Commissioner's Perspective," *Report of the Commissioner for the Environment and Sustainable Development* (2002). Similar comments regarding an implementation gap have been made in subsequent CESD reports.

6 Organization for Economic Cooperation and Development (OECD), OECD *Environmental Performance Reviews- Canada* (2004).

7 Ontario Ministry of Transportation, "McGuinty Government helps drivers go green," Media Release, 8 August 2007.

8 K.Harrison, "Retreat from Regulation: Evolution of the Canadian Environmental Regulatory Regime," in G.B. Doern, M.Hill, M. Prince, and R. Schultz, eds. *Changing the Rules: Canadian Regulatory Regimes and Institutions* (University of Toronto Press, 1999), 122–43.

9 M.Winfield, "The ultimate horizontal issue: the environmental policy experiences of Alberta and Ontario 1971–1992," *Canadian Journal of Political Science* 27.1 (March 1994).

10 Ibid.

11 M. Winfield and D. Macdonald, "Federalism and the Environment" in H.Bakvis and G.Skogstad, eds. Canadian Federalism: *Performance, Effectiveness and Legitimacy*, Second Edition (Oxford University Press, 2007).

12 "Commission on Conservation" *The Canadian Encyclopedia*. Available at: www.thecanadianencyclopedia.com/index.cfm?PgNm=TCE&Params=A1ARTA000178; accessed September 14, 2007).

13 P. Pross and R.S. Lambert, *Renewing Nature's Wealth: A Centennial History* (Ministry of Natural Resources, 1967) on the early development of forest legislation and policy in Ontario.

14 For a discussion of the "conservation" movement of the early 20th century see N.V.Nelles, "Conserving the Forest," *The Politics of Development: Forests, Mines and Hydro-electric Power in Ontario 1849–1941* (MacMillan of Canada, 1974), 182–214.

15 See www.oag-bvg.gc.ca/domino/cesd_cedd.nsf/html/cesd_index_e.html.

16 See www.eco.on.ca.

17 World Commission on Environment and Development (WCED), *Our Common Future* (Oxford University Press, 1987).

18 See www.iisd.org

19 Angus Reid Strategies, *National Political Landscape: Focus on the Environment*, January 23, 2007; Also see Decima Research, *Environment on the Agenda*, 4 January 2007.

20 McAllister Opinion Research, *The Environmental Monitor*, September 2006. See also McAllister Opinion Research *Global Warming A Telephone Survey of Canadians*, March 2008, which reports 42 percent of respondents identifying environment as the most important problem facing Canada today identifying climate change/global warming/GHGs as they key concern. Pollution in general was the next highest specific issue identified, being mentioned by 9 percent of respondents.

21 Data from Environics, *Focus Ontario* data retrieved from the CORA database (Queen's University, Fall 2007).

22 Environics, *The Environmental Monitor*, February 1996.

23 See The Hon.D.O'Connor, *Report of the Walkerton Inquiry: Part 1* (Queen's Printer, 2000), chapter 11.

24 See, classically, A.Downs, "Up and down with ecology: The issue attention cycle," *The Public Interest* 28 (Summer 1972), 38–50.

25 S.N. Soroka, *Agenda Setting Dynamics in Canada* (University of British Columbia Press, 2002).

26 A prominent example of the failure to expend announced funds related to programs for the restoration of the Great Lakes announced in 1994. See CESD, *Report of The Commissioner of Environment and Sustainable Development* (2002), chapter 1. Available at: www.oag-bvg.gc.ca/domino/reports.nsf/html/c2001menu_e.html; accessed 21 September 2007.

27 On the evolution and impact of the Green Plan see G. Toner, "The Green Plan: From Great Expectations to Eco-Backtracking … to Revitalization?" in Susan Phillips, ed. *How Ottawa Spends 1994–95: Making Change* (Carleton University Press, 1994), 229–60; G. Toner, "Environment Canada's Continuing Rollercoaster Ride," in Gene Swimmer, ed. *How Ottawa Spends 1996–97– Life Under the Knife* (Carleton University Press, 1996), 99–132.

28 Government of British Columbia, Backgrounder "BC's Revenue Neutral Carbon Tax." Available at: www.bcbudget.gov.bc.ca/2008/backgrounders/backgrounder_carbon_tax.htm; accessed 18 March 2008.

29 See J.Whitmore and C.Demerse, *Highlights of Provincial Greenhouse Gas Reduction Plans* (Pembina Institute, 2007). Available at: www.pembina.org/pub/1504; accessed 23 September 2007.

30 A.Taylor, M.Winfield, and M.Bramley, *Government Spending on Canada's Oil and Gas Industry: Undermining Canada's Kyoto Commitment* (The Pembina Institute, 2005).

31 Department of Finance, Budget 2007, *Preserving Our Environment and Modernizing Our Health Care System*. Available at: www.budget.gc.ca/2007/themes/papemhe.html; accessed 18 March 2008.

32 See Environment Canada, "Turning the Corner." Available at: www.ec.gc.ca/default.asp?lang=En&n=75038EBC-1; accessed 18 March 2008.

33 Government of British Columbia, *Backgrounder.*

34 Winfield, "The ultimate horizontal issue" (1994).

35 See D. Boyd, *Unnatural Law: Rethinking Canadian Environmental Law and Policy* (University of British Columbia Press, 2003), 239–240.

36 See for example, A. Kranjc, "Whither Ontario's Environment: Neo-Conservatism and the Decline of the Ministry of the Environment," *Canadian Public Policy* 26.1 (January 2000).

37 See Pross and Lambert on budgetary reductions to the Department of Lands and Forests during the Hepburn government in Ontario.

38 K.Harrison, *Passing the Buck: Federalism and Canadian Environmental Policy* (University of British Columbia Press, 1996), 58.

39 For 1989 and 2006 data see figure 2.

40 See, for example, S.Bennett, "Canadian Opinions on Environmental Policy: Patterns and Determinants, in A.Frizzell and J.H.Pammett, eds. *Shades of Green: Environmental Attitudes in Canada and Around the World* (Carleton University Press, 1997), table 2, 21. See also Environics, *The Environmental Monitor*, 1995–2000.

41 On the concept of punctuated equilibrium in evolutionary biology see Stephen Jay Gould, *The Structure of Evolutionary Theory* (Harvard University Press, 2002).

42 R.Mackie, "Health, education main Ontario issues, poll," *The Globe and Mail* 2 September 1999; The environment was identified as the issue which should receive the greatest attention of politicians in Ontario by 14 percent of respondents.

43 R.Mackie "Poll adds to troubles for Harris," *The Globe and Mail* 28 June 2001; R.Mackie, "Ontario Tories, NDP gain as Liberal Support Slides," *Globe and Mail* 2 July 2002.

44 Environics Research Group, "Ontario Environmental Coalition Ontario Voter Survey." Available at www.greenbelt.ca/reports/resultjuly07.pdf; accessed 5 November 2007. Both issues we identified by 18 percent of respondents.

45 J. McPhail, *The Health Effects of Ground Level Ozone* (Ontario Medical Association, May 1998).

46 S. Basrur, *Air Pollution Burden of Illness in Toronto* (Toronto Public Health, 2000).

47 O'Connor, *Report of the Walkerton Inquiry*, chapter 11.

48 On the number of smog advisories in Ontario since 1995, see www.airqualityontario.com/press/smog_advisories.cfm. The number of 'smog days' has risen significantly from 2000 onwards.

49 Up from 2.8% of popular vote in the 2003 election result.

50 Ipsos-Reid, "The McGuinty Government at its One Year Anniversary," October 2, 2004. Available at: www.ipsos-na.com/news/pressrelease.cfm?id=2390; accessed 5 November 2007.

51 Elections Ontario gives an unofficial figure of 8.0 percent of popular vote for the Green Party. Available at: www.elections.on.ca; accessed November 4, 2007.

52 J. Connelly and G, Smith, *Politics and the Environment: From Theory to Practice Second edition* (Routledge, 2003).

53 Ipsos Reid Public Affairs, "Ontario Provincial Election 2007 Exit Poll," 12 October 2007, table 21.

54 A collection of materials on the evolution of drinking water policy in Ontario in the aftermath of the Walkerton disaster is available on the website of the Canadian Environmental Law Association. Available at: http://www.cela.ca/coreprograms/detail.shtml?x=1437.

55 See M.Winfield, *Building Sustainable Urban Communities in Ontario: A Provincial Progress Report* (Pembina Institute, 2006).

56 See, for example, www.renewableisdoable.com; accessed 6 November 2007.

57 See M. Winfield, "Environmental Governance in Canada: From Regulatory Renaissance to Smart Regulation," *Journal of Environmental Law and Practice* 17.2 (2007), 69–83.

58 See Organization for Economic Cooperation and Development OECD *Environmental Performance Reviews: Canada*.

59 For a discussion of these critiques of "command and control" regulation see N. Gunningham, P. Grabosky, with D. Sinclair, *Smart regulation: Designing Environmental Policy* (Oxford University Press, 1998).

60 Government of Canada, *Cabinet Directive Streamlining Regulation* (Treasury Board Secretariat, 2007).

61 M. Winfield, "Environmental Governance in Canada: From Regulatory Renaissance to Smart Regulation," keynote address given at the annual Journal of Environmental Law and Practice Conference, Saskatchewan, June 2006.

62 Environment Canada, "Acid Rain: What's Being Done, What Has Canada Done?" Available at: www.ec.gc.ca/acidrain/done-canada.html; accessed 12 October 2006.

63 Environment Canada, "Implementing sustainable practices in the pulp and paper industry, A 10-year path to success" Backgrounder (June 2003). Available at: www.ec.gc.ca/press/2003/030606_b_e.htm; accessed 12 October 2006.

64 See for example, West Coast Environmental Law, *The BC Government: A One Year Environmental Review* (WCEL, 2002). Available at: www.wcel.org/wcelpub/2002/oneyearreview_final.pdf; accessed 6 November 2007. See also M. Winfield and G. Jenish, *Ontario's Environment and the Common Sense Revolution: A Four-Year Report* (Canadian Institute for Environmental Law and Policy, 1999).

65 See, for example, M. Walls, *EPR and Policies and Product Design: Economic Theory and Selected Case Studies* (OCED Working Group on Waste Prevention and Recycling, 2006)

66 R.Peters, M.Horn, and A.Ballie, *Successful Strategies for Energy Efficiency* (Pembina Institute, 2006).

67 See www.obc.mah.gov.on.ca/site4.aspx.

68 The regulations currently governing the sector were adopted in 1992 under the Canadian *Environmental Protection Act*, 1988 and the *Fisheries Act*.

69 CESD, "A Legacy Worth Protecting: Charting a Sustainable Course in the Great Lakes and St. Lawrence River Basin," *Report of the Commissioner of the Environment and Sustainable Development* (2001). Available at: www.oag-bvg.gc.ca/internet/English/oag-bvg_e_11662.html; accessed 18 March 2008.

70 S.Bernstein, "International institutions and the framing of domestic policies: The Kyoto Protocol and Canada's response to climate change," *Policy Sciences* 35 (2002), 203–36.

71 On recent developments in US climate change policy and the positions of the leading candidates in the 2008 presidential election see David B. Hunter, "The Future of US Climate Change Policy," in S. Bernstein, J. Brunnee, D.G. Duff, and A.J. Green, eds. *A Globally Integrated Climate Policy for Canada* (University of Toronto Press, 2008), 79–103.

72 See, for example, J. Rubin, "Coming home" in CIBC World Markets StrategEcon, 28 March 2008. Available at: http://research.cibcwm.com/economic_public/download/smar08.pdf; accessed 28 March 2008.

73 Green Ribbon Panel, *Fulfilling the Potential, A Review of the Environment and Sustainable Development Practice of the Office of the Auditor General of Canada* (Office of the

Auditor General of Canada, 2008). Available at: www.grp-gev.ca/reprap-e.htm#hd3b; accessed 18 March 2008.

74 Canada, The *Auditor General Act*, 1985 (c. A-17 as amended s.15.1).

75 Recyclable metals and plastics (blue box), paper and paper products (grey box), household organics (green bin), leaf and yard wastes, household hazardous wastes, residual wastes, and, in some households, disposable diapers.

76 City of Toronto, "Residential Diversion Rate." Available at: www.toronto.ca/garbage/residential-diversion.htm; accessed 24 October 2007.

6 The Politics of Sustainability in a Complex Federal State

ROGER GIBBINS

Although climate change is not a new public policy issue,[1] the recent popularity of *An Inconvenient Truth*, aggressive policy initiatives such as California's low carbon fuel standard, and the ongoing debate about emission reduction targets and how to meet them have made it one of the most important public policy topics of the day. Indeed, governments across Canada and around the world are exploring a wide range of policy responses to growing public concern. Admittedly, debate over the science of climate change and the impact of human activity on global warming continues, but this debate is essentially moot from a public policy perspective.[2] Governments everywhere are taking steps to alter human activity in response to public demand for action on climate change, and Canada is no exception to this international trend. However, global warming presents a difficult policy challenge for democratic states, and a particularly difficult challenge for federal states. In the Canadian case, it is rendered even more difficult by the country's regional diversity.

My intent here is to explore how environmental issues generally, and global warming more specifically, are handled within the Canadian federal state. The basic argument is that the interaction between Canada's vast geography and federal form of government complicates the environmental policy file, and that these complications are brought into bold relief when we turn from environmental matters writ large to the specific issue of global warming. Furthermore, I will suggest in conclusion that global warming also constitutes a potential challenge to the Canadian federation, one that may exacerbate regional tensions and, perhaps, transform federalism as we know it.

Given that discussions of climate change and global warming have become so polarized and intensely ideological in character, it might also be useful to state at the outset where I stand. Simply put, climate change is a global challenge that requires a proactive and creative public policy response by all Canadians and their governments. Central to this response is changing the way we produce and consume energy, and it is this reality that thrusts western Canada to the centre of the national policy debate, and hopefully to a position of national leadership.

ENVIRONMENTAL POLITICS IN FEDERAL STATES?

Environmental issues *per se* are not intrinsically difficult for federal political systems to handle. Indeed, the relatively decentralized character of federal states may be well suited for the localized character of many environmental challenges; the mantra to "think globally and act locally" aligns nicely with federal systems of government. It should not be surprising, therefore, that Canada has done a reasonably good job of environmental management, although perhaps our record does not match the smugness that we so often bring to the file, particularly when making ill-informed comparisons with the United States. Virtually all governments – federal, provincial, and municipal – have comprehensive environmental programs in place, and those governments have handled a complex array of environmental challenges ranging from acid rain to wildlife protection. Now admittedly, one of the reasons that the Canadian environment is in pretty good shape is that we have a lot of it; it is difficult for 33 million people to have too destructive an impact upon such a huge chunk of real estate. Nonetheless, we have not done a bad job. It is also clear, however, that global warming presents a policy challenge of an entirely new magnitude. Thus our success to date may not predict how well the Canadian political system will handle the challenge of climate change.

But why will global warming/climate change be such a difficult problem for the Canadian federal state? Part of the difficulty comes not from the federal features of our political system but from the democratic underpinnings of the Canadian federal state. Global warming demands that governments impose costs on the current electorate while the compensating benefits will not accrue for generations to come.[3] In effect, Canadians will be asked to pay now, and collect much, much later; it is likely to be their great grandchildren who will reap the benefits. Moreover, while the costs of current climate change initiatives are localized (Canadian taxpayers), the beneficiaries are not only distant but very diffuse (the world, humankind). And, many of the beneficiaries are not even voters no matter how far ahead we might look; here the ubiquitous polar bear on his ice floe comes to mind. This is not to say that we are incapable of acting on behalf of non-voters, as the new Species at Risk Act demonstrates. Our engagement in Afghanistan and foreign aid budget also show our willingness to act in the interest of non-Canadian voters, although declining public support for the former and the modest nature of the latter speak to the very real limits on Canadian altruism.

Global warming forces governments to construct policy architectures that stretch well beyond the three to four year electoral cycles, or even shorter cycles in the case of minority governments, that provide the conventional boundaries for policy debate. For example, the year 2050, which looms large in contemporary policy debates about GHG emission targets, is at least 15 governments away in the Canadian case.[4] As a consequence, there are huge incentives for inaction, for passing the global warming buck to governments yet to come, or for substituting rhetoric for action, which explains in part why the Government of Canada did so little after signing onto the Kyoto Accord.

In a similar fashion, "creeping crises" are difficult for democratic systems to handle. Note, for example, our ongoing financial struggle with infrastructure maintenance.[5] If we cannot see the inflection point where the distant crisis suddenly becomes immediate, it is difficult to mobilize the political will and resources to act.[6] Therefore democratic states face a difficult challenge, which is not to say that less democratic states are leading the environmental charge; deciding when future conditions demand action today is tough for any government. The democratic political process works best when incremental change is adequate. At issue, then, is whether incremental change is a sufficient response to the climate change challenge, although I suspect that incrementalism will not take us where we need to go.

Finally, global warming poses a significant public administration challenge to democratic states. Simply put, it does not fit neatly into siloed departments. Yes, it is an environmental issue, but it is also a finance issue, an energy issue, an industrial policy issue, a transportation and health care issue, and quite possibly a determining issue for foreign aid and foreign policy. The challenge is to find a public administration, and therefore cabinet model, that fits somewhere between a climate change tsar, at one extreme, and broadly diffused responsibilities that impair policy coherence and accountability, at the other extreme.

Federalism itself comes into play when the focus of environmental concern shifts from local issues for which local action makes sense to issues that sweep across political lines on maps, and across jurisdictional boundaries within the federal state. The territorial fragmentation of governance within federal states can be problematic when the issues under debate transcend jurisdictional borders, as many water and airshed issues are prone to do. And here, of course, *global* warming is the ultimate example of an environmental issue that transcends not only lines on maps within federal states but also national boundaries around the world. If national boundaries and national sovereignties complicate a global response, then the complications are further multiplied within federal states with reasonably robust and autonomous sub-national governments, as is the Canadian case.

The complexities arising from jurisdictional boundaries, from the constitutional division of powers characteristic of federal states, come into play when we recognize that an effective climate change strategy cannot be constructed under a single head of power. As noted above, it is not strictly an environmental issue, or a finance issue, or an energy issue. The situation becomes even messier because the relevant constitutional powers reside with both the federal and provincial/territorial governments. Both orders have environmental portfolios, both deal with energy, both deal with infrastructure and urban affairs, and both have an extensive set of tax penalties and incentives that can be yoked to environment action. This general messiness makes it more difficult to address policy design through conventional intergovernmental mechanisms; deciding who should be at the table, or for that matter, what the table/s should be will be difficult. We know, moreover, that even though intergovernmental approaches and solutions will be needed, they will not be easy. Executive federalism works best in Canada when provincial and territorial leaders sit down to divide up new pools of federal money; policy design apart from dividing up the spoils does not come easily to those leaders, as the prolonged national debate over health care reform demonstrates so vividly.

The reality is that any true *national* climate change strategy or energy strategy will necessarily be a complex and messy amalgam of federal, provincial, territorial and big city frameworks. The price of federalism may often be a lack of policy coherence. However, and on a more positive note, federalism creates space for policy innovation and for state/provincial policy leadership when national governments are unable or unwilling to act. State and provincial governments can serve as important policy laboratories. It is California, for example, rather than Washington, DC, that set the climate change policy pace in the United States prior to Barack Obama's election, just as British Columbia rather than Ottawa set the pace in Canada, at least prior to Stéphane Dion's Green Shift initiative.

In summary, global warming is different in character because it is not localized in the way that so many environmental issues were in the past. As a consequence, the mantra to "think globally, act locally" does not apply; Canadians appear to be demanding not only a local/provincial policy response but also a national and international policy response. However, the demand for, or at least public expectations for, a national policy response takes us to the nub of the Canadian dilemma, and that is how to construct a national response within the context of a regionally diverse federal state with its divided jurisdictions. Federalism institutionalizes regional interests, and these regional interests are brought to the fore in the global warming policy debate. This point is so important that it warrants a more detailed discussion.

REGIONAL DIMENSIONS OF CLIMATE CHANGE

It would be unfair to say that *all* public policy issues in Canada are entangled in regionalism, although the fact that support for Canada's Afghanistan mission breaks so sharply along regional lines shows the pervasive potential of regionalism to shape and complicate the national public policy agenda. However, in the case of environmental politics generally, and in the specific case of climate change, there is no question that regionalism has the potential to greatly complicate the policy debate.

While all countries face both perceived risks associated with climate change and political imperatives to respond, the impact of climate change will not play out evenly across countries or across regions within vast transcontinental countries such as Canada. The prairies, for example, may face exposure to drought conditions that other Canadian regions will not face, whereas climate change in British Columbia could transform timber harvests to a degree that other provinces will not encounter. Thus while all countries face climate change policy challenges as a green wave sweeps across the political landscape, *the Canadian policy debate is greatly complicated by regional diversity* that goes well beyond differences in physical geography:

- The industrial structure of the Canadian economy varies substantially across regional communities, and as a consequence neither climate change nor climate change policies will play out evenly across our loosely integrated national economy. It is difficult, for example, to imagine a one-size-fits-all policy framework that could be applied to the auto industry in Ontario and the oil and gas industries in western Canada.
- Canada's population and economic growth are unevenly distributed, and therefore the implications of climate change policies for population and economic growth are also unevenly distributed. Regions with sustained economic and population growth face a more difficult challenge than do low-growth regions.
- Because the effects of climate change will vary dramatically across regions, we will lack unifying national symbols around which popular support for a national policy response might be built; there are no Canadian analogues to the melting glaciers in Switzerland, or the pervasive impact of drought in Australia.

These regional differences make a *national* policy response to climate change very difficult. The policy challenge is then multiplied many times when we layer on regional differences in the distribution and mix of energy resources. Canada's vast energy resources, including conventional oil and gas, coal, hydro, uranium, the oil sands and potentially biofuels, are not evenly distributed across the country. They bunch up in ways that create dramatic differences in circumstances among the ten provinces and three territories. Provinces also differ greatly in the centrality of energy resources to their provincial economies, and in the relative importance of energy exports and imports. There are even significant regional differences in the mix of energy supplies and consumption patterns. This variability is then reinforced, institutionalized, and politicized by the fact that Canada is a federal state, with the ownership and management of natural resources resting largely with provincial governments.

But why are these regional differences in energy resources, supply, trade and consumption so important in the context of potential policy responses to global warming/climate change? The answer is simple: all such policy responses entail, in one way or another, moving towards a carbon-constrained future with respect to the production and consumption of energy. If climate change is the problem, solutions are sought in how we produce and consume energy. This is the dominant policy interface; everything else is secondary. Thus when the national debate on global warming slides into a national debate on energy strategies, as it inevitably will, Canadians will face an extraordinarily complex policy challenge given the deep basket of regional differences described above. This is not to suggest that coherent and effective national policies be subordinated to regional interests, but rather to recognize that regional differences matter, and that they matter here more than they do in virtually any other policy domain, and that how they are handled is of great consequence for the success of climate change policies, and for the health of the federation.

If our political history teaches us anything, it is that national policy initiatives that touch upon the provincial ownership of natural resources can be politically charged and highly contentious. Here we need only note the impact of the 1980 National Energy Program on the national political landscape, the inability of the Council of the Federation to agree in August 2007 to even the rough outlines of a national climate change policy response, the warning of former Alberta Premier Peter Lougheed about the explosive potential of a constitutional battle over environmental legislation impinging on Alberta's oilsands, and the wariness that many western Canadians expressed in the 2008 general election campaign regarding the proposed Green Shift.[7] We know from our history how contentious policies that affect energy resources and production can be, and how deep the regional fault lines can run. The risks, moreover, extend beyond the West to the rest of the federation, for how the energy-rich West responds to the climate change challenge is truly of national significance. Therefore, both the West and the national community within which it is embedded have an important stake in getting the policy architecture right by addressing climate change in a way that moderates rather than exacerbates regional tensions.

Now admittedly, energy policies are only a subset of a more comprehensive public policy response to climate change and global warming. However, they lie at the very core of that response; they are the things we have to get right if we are to have any hope of succeeding. We are faced, then, with a policy challenge that is not only complex but also terribly important for Canada's future: *how do we construct a national energy strategy for a carbon-constrained future, and how do we do so without exacerbating regional tensions within the federation?*[8]

Those who say regional interests be damned, that we have bigger global fish to fry, are running a serious risk. They are assuming, for instance, that Quebecers will readily put aside generations of work in defending their province's jurisdictional autonomy, or that Albertans will identify more intensely with a global interest than with their provincial community. Nothing in the policy debate to date gives credence to such assumptions. Moreover, it would be foolish to assume that national policies will be effective if they stir up regional conflict or debates about the position of Quebec within Canada. Therefore figuring out how to act nationally within the context of a regionally diverse federal state is not just important for the health of the federation; it is also important if we are concerned about the effectiveness of policy frameworks.

THE MEDIATING ROLE OF PARTY POLITICS

Governments in democratic countries, and certainly in the Canadian federal state, are animated by party politics. It is not the Government of Canada that acts, or the Government of Ontario, but the *Conservative* Government of Canada and the *Liberal* Government of Ontario. But why does this simple observation matter with respect to Canada's policy response to global warming and climate change?

The primary importance of parties, and the partisan composition of the federal government, does not stem from differences in party platforms. Admittedly, these differences exist; the Conservatives, Liberals, New Democrats and Greens do not see eye-to-eye on global warming, but then neither do the Liberals-in-government and the Liberals-in-opposition, or for that matter the early 2006 Conservatives and the late 2007 Conservatives. However, these often ephemeral differences in party platforms pale beside structural differences in regional support.

Partisan support in Canada is deeply fragmented along regional lines. The Bloc's support is, of course, restricted to Quebec, and therefore the Bloc has no interest in *Canadian* policy towards global warming other than whether it will convey benefits for or inflict costs on Quebec. The Liberal party is increasingly a regional party based in southern Ontario, albeit with residual pockets of support in Atlantic Canada, anglophone Montreal, and Vancouver. There is a risk, then, that the party will champion a global warming strategy that works for southern Ontario rather than a strategy that just works, period. Conservative policy preferences are constrained in turn by their strong electoral base in the West, although the Conservatives are also the numerically dominant party in Ontario.

These regional differences in partisan support reflect and reinforce the regional differences in energy resources discussed above. The absence of truly national parties makes the construction of a truly national global warming strategy that much more difficult. In short, *who* is in power nationally counts a great deal in this particular policy domain; political parties anchored in specific regional communities are unlikely to gravitate towards a policy consensus because those communities are themselves so different. Brokering a national deal with the requisite regional tradeoffs requires national brokers, and these are chronically absent.

A regionally segmented party system increases the odds that climate change policies will morph into redistributive policies, that we will be more concerned about the regional distribution of costs and benefits than with whether policies are effective in meeting climate change goals. For example, the Government of Canada, under the guise of addressing climate change, may seek to restore a more equitable distribution of wealth even at the risk of a loss of efficiency in meeting climate change objectives.[9] Different partisan configurations could differ significantly on where to place the costs of emission reductions; should those costs be borne by energy producers in the West or energy consumers across the country? Should public investment to reduce GHG emissions be spread thinly across the country or directed to those areas where the traction on emissions might be greatest?

Now, it may be that the federal government, regardless of its partisan composition, will keep its eye on the global warming policy ball, and will let the regional chips fall where they may. More realistically, however, a regionally segmented and borderline dysfunctional national party system poses a major obstacle to the construction of an effective *national* policy response to global warming. It increases the risk that we will direct our policy guns at the redistribution of wealth rather than at the issue at hand.

OTHER POLITICAL CHALLENGES

This discussion of a regionally segmented party system does not exhaust the political problems likely to arise as we try to construct a national policy response to the challenges of climate change and global warming:

- The level of public support for aggressive action with respect to climate change is uncertain, and has not yet been tested by tough economic times. Here it is important to note public resistance to increases in the price of gasoline, in the price of electricity, in pricing at all when it comes to water.

- Although there are always international influences on Canadian domestic politics, the global warming debate has been internationalized, globalized, from the get-go. We are seeing a domestic debate about a global issue, indeed one that has been linked to the very survival of life as we know it. As a consequence, we are likely to see the incursion of international negotiations, Environmental Non-governmental organizations (ENGOs), and American movie stars. The domestic policy debate will not be contained within Canada, and will be driven by international targets; it will really be an international debate taking place *within* Canada.
- Environmental politics is an area where external policy expertise may be greater than internal capacity within the public sector. It will not be easy for governments to keep up.
- Canada's global warming challenge is complicated by the fact that Canada is a net energy exporter, although the form and benefits of exports differ dramatically across the country. Canada's relatively small population will never provide an adequate market for the country's huge energy assets.

In drawing this discussion of political problems to a close I should note that the difficulty the national government may have does not preclude successful action by provincial and local governments. In the United States, for example, state governments moved aggressively despite the lack of movement in Washington during the administrations of George Bush. The very nature of Canada may make it difficult for Ottawa to act, but these difficulties may not preclude effective action elsewhere. At issue, then, is whether thinking globally but acting locally – through provincial and municipal governments – will be a sufficient response given the magnitude of the challenge, or whether the federal government will have to get its act together.

A CHALLENGE TO THE FEDERATION?

If Canadians are not smart in their policy response to global warming and climate change, if we craft a policy response that is designed more to redistribute wealth than it is to reduce GHG emissions, we will fail our international obligations, and ourselves. However, the risks go well beyond policy failure. A redistributive policy response will enflame regional conflict and, as a consequence, could cripple any effective national policy response to global warming and climate change. Of course, there is nothing new here; regional differences and regional conflict can complicate even the most benign policy file. In this case, the risk to our national policy capacity, and for that matter to the federation, is simply much greater as an ill-conceived national energy strategy driven by a regionally fragmented party system could rupture the federation.

There is also another potential although not unrelated impact on the federation, and on the nature of federal governance. *If* global warming climbs to the top of the Canadian policy agenda and sticks, and *if* Canadians become even more insistent that their governments react aggressively to deal with climate change issues, then global warming may tip the federal balance of power in favour of the national government. Although provincial policies will always be an important part of the overall national response, Canadians may demand more sweeping action from their federal government. We can see hints of this potential in Australia where a prolonged drought is prompting an increasingly expansive policy response by the Commonwealth government, and we have ample evidence of how federal states centralize during times of war when a paramount national interest sweeps away jurisdictional boundaries. The risk in the Canadian case is that this swing to the national government could be accompanied by global warming policies shaped by redistributive objectives and regional insensitivity. If a more centralized federal system is the price to be paid for an effective climate change strategy, then we have to make sure that we get the design of that new federal order right.

NOTES

1 The Canada West Foundation, for example, released a report on the subject in 1994 entitled *The Climate for Debate: Global Warming and Policy Instruments for Emission Reduction.*

2 The terms *climate change* and *global warming* are often used interchangeably in Canadian political debate, particularly because the former is portrayed as the manifestation or consequence of the latter. I will tend to use global warming as the dominant term, but will slip into the climate change terminology when it seems more descriptive of the issues at hand, or for stylistic variation.

3 The assumption that climate change policies will be cost-free to the current electorate, which we can get by with a bit more regulation here and there, is not one that should be seriously entertained.

4 To put this into perspective, it is as if the global warming policies for today were crafted during the John Diefenbaker/Lester Pearson minority government period in the early 1960s. Would they have had the prescience to see this far ahead? Certainly global warming was not on their radar. Thus to behave today as if we can understand and predict the policy landscape in 2050 is the height of hubris.

5 We did, however, address concerns about the financial sustainability of the Canada Pension Plan.

6 Hence the appeal of Al Gore's hockey stick model, which provides and dramatizes an inflection point.

7 In a 14 August 2007 Calgary address to the annual meeting of the Canadian Bar Association, former Alberta Premier Peter Lougheed argued that pressure on the federal government for strong environmental legislation to reduce green-house gas emissions could threaten Canadian unity. "Alberta ground zero for green battle," *Calgary Herald*, 15 August 2007, A1.

8 This issue is addressed at length in Roger Gibbins and Kari Roberts, *Canada's Power Play* (Canada West Foundation, September 2008).

9 It is impossible not to reference the 1980 National Energy Program, which used the pretence of energy policies to redistribute wealth and economic opportunities, although to be fair much of the redistributive intent was directed towards federal coffers.

7 Education for Sustainable Development: Cure or Placebo?[1]

DAVID V. J. BELL

WHERE HAVE WE BEEN?
THE EMERGENCE OF THE CONCEPT OF ESD

Our Common Future, the Report of the World Commission on Environment and Development (more commonly known as the Brundtland Report), was *a* – perhaps *the* – seminal document in the emergence of sustainable development discourse, policies and practices. Despite its comprehensive discussion of SD issues and challenges, the *Report* devotes relatively little attention to education. Noting the imbalance between developed and developing countries *Our Common Future* calls for increased literacy overall and reduced gaps between male and female primary education enrolment rates. Linked to this concern for extending basic education globally is a brief discussion of the importance of improved environmental education, and a prescient call for a new kind of education that foreshadows (without using the term) the concept of Education for Sustainable Development (ESD): "Education should therefore provide comprehensive knowledge, encompassing and cutting across the social and natural sciences and the humanities, thus providing insights on the interaction between natural and human resources, between development and environment."[2]

Perhaps the most powerful statement in *Our Common Future* about the importance of education came from Gro Harlem Brundtland's foreword, where she underlined the importance of communicating a "message of urgency" to "parents and decision makers" in a language "that can reach the minds and hearts of people young and old." Ultimately, she insisted, "The changes in attitudes, in social values, and in aspirations the report urges will depend on vast campaigns of education, debate, and public participation."[3]

Our Common Future was officially a report to the United Nations, which responded by organizing UNCED: the United Nations Conference on Environment and Development held in Rio de Janeiro in June 1992. At UNCED the concept of SD was operationalized in a number of documents, the most comprehensive of which was *Agenda 21*. In the lead-up to UNCED many became convinced that if SD were to succeed it would require supportive education. Ultimately this resulted in the development of the concept of ESD. Indeed the term education appears in every chapter of *Agenda 21* and is used throughout the document with less frequency only than the term government. More importantly, education was given its own chapter (chapter 36) entitled "Promoting Education, Public Awareness and Training."

The initial pressure to develop ESD came from outside the education community, from international organizations, from business, governments and NGO's. To some members of the chapter 36 writing team,[4] all that was needed was more emphasis on environmental and outdoor education. Others pointed out that in some parts of the world, formal education was available only to a minority of young people and often for very short periods of time. (In parts of Latin America and the Caribbean, for example, the average level of educational attainment is Grade 4. In parts of Africa it is measured in months.) Given this stark reality, chapter 36 begins by recommending the preparation of national strategies and actions for meeting basic learning needs, universalizing access and promoting equity, broadening the means and scope of education, developing a supporting policy context, mobilizing resources and strengthening international cooperation to redress existing economic, social and gender disparities which interfere with these aims. Only after endorsing this plea for improved basic education (which arose from the World Conference on Education for All: Meeting Basic Learning Needs, Jomtien, Thailand, 5–9 March 1990) does the chapter address the content of ESD by calling for "strategies aimed at integrating environment and development as a cross-cutting issue into education at all levels."[5]

Chapter 36 embraced a broad definition which included education offered in classrooms (formal education) as well as in non-school settings such as workplaces or religious organizations (non-formal education) and the broader forms of communication that help shape public awareness and attitudes (informal education). In short, it identified four major thrusts to ESD:

1 improve basic education (especially in the developing countries)
2 reorient existing education to address SD
3 develop public understanding and awareness
4 provide training for all sectors of society including business, industry and government.

Even though the concept of ESD and its application have evolved over the past 15 years, many of the tenets set forth in *Agenda 21* have proved seminal. These key elements include:

- the broad definition noted above (formal, non-formal, informal)
- the recognition that ESD must take into consideration the local environmental, economic, and societal conditions. As a result ESD will take many forms around the world
- the insistence that ESD is education for (not about) sustainable development[6]
- the notion of linking knowledge, values, perspectives, and skills/behaviour (the head, the heart, the hands)
- recognition of the importance of aboriginal and traditional knowledge
- the importance of supportive educational policy
- the need for various forms of teacher training and professional development for educational administrators and other key education decision makers

The imperative of teacher training might have posed huge obstacles to the efforts to promote ESD. How could resources be found to present entirely new content and pedagogy to the world's 55 million [now over 60 million] teachers? The response was a clever reframing and wise adoption of what was termed a "strengths approach". Instead of assuming that teachers' sustainability knowledge glass was basically empty and extensive new training would be needed to remediate this deficiency, the strengths approach recognized that much of what needs to be taught is already in the curriculum (though not identified explicitly as part of ESD), and that all teachers can contribute to sustainability education once they are made aware of sustainability issues and perspectives: "Rather than primarily retraining inservice teachers to teach sustainability, we need to design new approaches to pre-service and inservice teacher education to address Environment Economy Society."[7]

ESD learning objectives connect to each and every discipline and subject:

No one discipline can or should claim ownership of ESD. In fact, ESD poses such broad and encompassing challenges that it requires contributions from many disciplines. For example, consider these disciplinary contributions to ESD:

- Mathematics helps students understand extremely small numbers (e.g., parts per hundred, thousand, or million), which allows them to interpret pollution data.
- Language Arts, especially media literacy, creates knowledgeable consumers who can analyze the messages of corporate advertisers and see beyond "green wash."
- History teaches the concept of global change, while helping students to recognize that change has occurred for centuries.
- Reading develops the ability to distinguish between fact and opinion and helps students become critical readers of political campaign literature.
- Social Studies helps students to understand ethnocentrism, racism, and gender inequity as well as to recognize how these are expressed in the surrounding community and nations worldwide.[8]

Chapter 36 outlined both deadlines for accomplishing its key recommendations and funding estimates to support them. However, as with other efforts to secure funding for *Agenda 21*, money was never forthcoming. Deadlines came and passed. But the impetus from Rio continued to energize and inspire ESD developments around the world.

ESD AS A PROCESS WITH A PURPOSE

Many sustainability theorists and practitioners agree that sustainable development is a process rather than an outcome – a process that features ongoing learning and "adaptive management." Not surprisingly, similar views have been put forward about ESD: "We believe that education for sustainability is a process, which is relevant to all people and that, like sustainable development itself, it is a process rather than a fixed goal. It may precede – and it will always accompany – the building of relationships between individuals, groups and their environment. All people, we believe, are capable of being educators and learners in pursuit of sustainability."[9]

But what purpose guides this process? This question has generated considerable controversy, largely focused on whether ESD is education *about* or *for* sustainable development:

An important distinction is the difference between education *about* sustainable development and education *for* sustainable development. The first is an awareness lesson or theoretical discussion. The second is the use of education as a tool to achieve sustainability ... While some people argue that "for" indicates indoctrination, we think "for" indicates a purpose. All education serves a purpose or society would not invest in it ... ESD promises to make the world more livable for this and future generations. Of course, a few will abuse or distort ESD and turn it into indoctrination. This would be antithetical to the nature of ESD, which, in fact, calls for giving people knowledge and skills for lifelong learning to help them find new solutions to their environmental, economic, and social issues.[10]

ESD IN CANADA

While the origins of ESD can be traced to *Agenda 21*,[11] even before the Earth Summit and the adoption of *Agenda 21*, work had begun in Canada to evolve the concept of ESD. These efforts resulted (indirectly) from the work of the Brundtland Commission which in 1986 had held a series of eight public hearings in Canada. The impact was profound. Canada's response to the work of the Commission (whose secretary general was Canadian Jim MacNeill, with Maurice Strong a prominent member) included establishing (in 1986) a "National Task Force on the Environment and the Economy" (NTFEE) whose recommendations led (*inter alia*) to the later establishment at the national, provincial, and local levels, of "Round Tables on the Environment and the Economy." Among the other fruits of Brundtland and NTFEE was the founding of the International Institute for Sustainable Development (IISD) in Winnipeg, and the strong support for SD from the Government of Manitoba which had arguably the most effective provincial Round Table in Canada, and also the first SD Act, passed in 1997.[12]

It was the National Round Table on the Environment and Economy (NRTEE) which saw most acutely the need for changes to the education system in Canada. In 1991, NRTEE helped set up an NGO charged with the task of promoting sustainability education across the curriculum in elementary and secondary education in Canada. The first Chair of Learning for a Sustainable Future (LSF) was a former CEO of Shell Canada, Jack McLeod, who apparently was personally encouraged to take on this mission by Jim MacNeill. The founding executive director was Jean Perras. Unlike NRTEE, which had always received financial support from the federal government and later became more permanently established through an Act of Parliament, LSF from the outset was required to raise its own funding from a variety of sources including governments, businesses, and foundations.

Perras began his work by conducting consultations across the country with key education stakeholders in an effort to develop a consensus on the appropriate content and pedagogy for ESD. Both Hopkins and Perras were reminded that some of the key building blocks for ESD were already in place in such concepts as Environmental Education and Global Education. Indeed there were some who objected to ESD as either redundant or wrong-headed.[13]

The task of any Canadian national organization set up to promote educational objectives is complicated by the constitutional division of powers which (under Section 92) assigns full responsibility for (formal) education to the provincial governments. As a national organization working in an area of provincial responsibility, to be effective LSF would need to develop good working relationships with provincial ministries of education and related teacher and school board organizations. The LSF Board drew its membership from all regions of Canada, but as one might expect it had more success in some provinces than in others. In most instances LSF worked through loose alliances or partnerships, but Ontario followed a different path. With support from the Ontario Round Table on the Environment and the Economy (ORTEE), various Ontario ministries, NRTEE, and with guidance from LSF, the Ontario Learning for Sustainability Partnership (OLSP) was established in 1994 to promote ESD across the province. The initiative was very nearly stillborn when the election in June 1995 brought to power the Mike Harris Conservatives. Within weeks the Premier's Council and ORTEE were disbanded, thus removing the principal institutional supports for OLSP. The Harris Conservative government followed this move up a year later by taking environmental education out of the Ontario curriculum. OLSP managed to survive, however, and soon strengthened its institutional links to LSF, renaming itself LSF-O. Several years later, after Jean Perras had stepped down as executive director, the two organizations merged and the executive director of LSF-O, Pam Schwartzberg, was appointed to head the national organization.

The LSF strategy has focused on advancing (ESD) policy in Canada; supporting educators and youth through workshops and resources; and working with students, teachers, and business and community organizations to facilitate sustainability action projects. This latter strategy has been accomplished primarily through a series of Youth Taking Action Forums which were pioneered in Ontario but now have been held in all regions of Canada. This reflects LSF's mission to promote the knowledge, skills, perspectives and practices essential to a sustainable future by helping citizens acquire, through education, the knowledge, skills, perspectives and practices needed to contribute to the development of a socially, environmentally, and economically sustainable society not only for today but for generations to come. LSF works with educators, students, parents, government, community, and business to integrate the concepts and principles of sustainable development into education policy, school curricula, teacher education, and lifelong learning across Canada and beyond.[14]

As a follow-on to Rio and Agenda 21, in October 1992, Chuck Hopkins organized the "Eco-ED" conference in Toronto which attracted nearly 6000 delegates (including more that 500 aboriginal delegates) from all parts of the world. The heads of six UN agencies attended, and each panel featured a participant from government or education, a business representative, an aboriginal person, and at least one woman. Whereas Hopkins has worked primarily with UNESCO (and later with UNU), Perras was strongly linked to the International Union for the Conservation of Nature (IUCN).[15] Both groups co-sponsored a second large international Conference on ESD in Thessaloniki, Greece in 1997. Hopkins convened a workshop on Teacher Education for Sustainability (TEFS) that brought together leading educators from a number of countries to strengthen TEFS globally, particularly in the developing countries. The Thessaloniki Declaration clarified the scope and reach of ESD by pointing out that: "The concept of sustainability encompasses not only environment, but also poverty, population, health, food security, democracy, human rights and peace. Sustainability is, in the final analysis, a moral and ethical imperative in which cultural diversity and traditional knowledge need to be respected."[16]

Many organizations have articulated the implications of this broad construction of sustainability for ESD. The following excerpt from a Quebec document is particularly eloquent:

Education for a sustainable future can therefore be seen as "a concept that is wider than the environment. It rejects durable development [i.e. SD], which is considered to be too vague. It also rejects the original concept of environmental education, which is seen as too narrowly linked to the natural environment. Education for a sustainable future seeks to integrate these two concepts into other, wider concepts: non-violence, peace, co-operation, human rights, democracy ...

Whereas environmental education, as originally defined, remains committed to maintaining its close links with the environment, education for a sustainable future seeks, in contrast, to become a horizon for integrating other educational currents. This is clearly not a particular subject that should take a place alongside other subjects. Instead, it seems to fit into the field of transversal skills discussed in the education policy statement of the Ministère de l'Éducation (1997), while at the same time being open to a wide variety of educational subjects."[17]

WHERE ARE WE NOW?
UNDESD AND CANADA'S RESPONSE

In 2004, the UN declared 2005 – 2014 the UN Decade of Education for Sustainable Development (UNDESD). In proclaiming the Decade, UNESCO identified its purpose as follows:

The goal of the United Nations Decade of Education for Sustainable Development (2005–2014, DESD), for which UNESCO is the lead agency, is to integrate the principles, values, and practices of sustainable development into all aspects of education and learning. This educational effort will encourage changes in behaviour that will create a more sustainable future in terms of environmental integrity, economic viability, and a just society for present and future generations.[18]

This proclamation provided a renewed emphasis on ESD around the world, and sparked a number of important initiatives here. Canada signaled its support for the Decade at a high-level meeting of Environment and Education Ministries in March 2005. The purpose of the meeting was to adopt the Economic Commission for Europe's (UNECE) Strategy and Implementation Framework for the Decade. Gerald Farthing, deputy minister, Manitoba Education, Citizenship and Youth led the Canadian Delegation with Suzan Bowser, director general, Environment Canada;[19] Diane Rochon, program officer, Ministère de l'Éducation Québec;[20] and David Walden, secretary-general, Canadian Commission for UNESCO.

The UNECE Strategy encourages UNECE member states to develop and incorporate ESD into their formal education systems, in all relevant subjects, and in non- formal and informal education. The *aim* is to equip people with knowledge of and skills in sustainable development, making them more competent and confident and increasing their opportunities for acting for a healthy and productive life in harmony with nature and with concerns for social values, gender equity and cultural diversity. The *objectives* of the UNECE strategy are to:

- ensure that policy, regulations and operational frameworks support ESD
- promote ESD though formal, non-formal and informal learning
- equip educators with the competence to include ESD in their teaching
- ensure that adequate tools and materials for ESD are accessible
- promote research on and development of ESD
- Strengthen cooperation on ESD at all levels.
- Foster conservation, use, and promotion of knowledge of Indigenous Peoples in ESD.

The UNECE strategy provides member states with an implementation plan and a set of indicators to report progress toward meeting the objectives. CMEC, with the help of the Canadian Commission for UNESCO, agreed to take on the task of reporting to the UN on the implementation of the UNECE Strategy using this framework and indicators.[21] Gerald Farthing is serving as the CMEC lead on this project.

To help deepen Canada's response to UNDESD, in 2005 LSF initiated a three way partnership (federal government, provincial government, and NGO) to advance ESD in Canada. Partners include Environment Canada, Manitoba Education, Citizenship and Youth (MECY)/ Manitoba Advanced Education and Literacy (MAEL), and LSF. Together these three organizations comprise a steering committee for an ambitious series of initiatives coordinated by LSF in support of the UNECE objectives:

1 Provincial/territorial Education for Sustainable Development Working Groups have been established to date in eight jurisdictions in Canada.
2 A National Education for Sustainable Development Expert Council (NESDEC) has been formed.
3 The Canadian Sustainability Curriculum Review Initiative is in progress.
4 The ESD Resource Database ("Resources for Rethinking – R4R") has been launched.
5 Youth Taking Action Forums (which include Teacher Professional Development Workshops) are being held in several locations across the country.

Following from the first of these initiatives Manitoba established a pilot ESD provincial working group in 2005. Similar groups have since been set up in BC, Alberta, Saskatchewan, Ontario, New Brunswick, Nova Scotia, PEI, and Newfoundland. Although Quebec has opted, to date, not to form a Working Group, there is a strong commitment to ESD flowing from the 2006 Sustainable Development Act (see Appendix A[22]) and the work of the CSQ which enthusiastically endorsed Education For Sustainability (EFS) in the 1990s, and has given continuing support to the Ecoles Vertes Brundtland (Brundtland Green Schools) of which there are now over 900 in the province.

Despite only minimal funding from the federal government ($15k per year for 2006–08, and a recent decision not to extend funding beyond March

2009) many of these working groups have been very active and remarkably successful, as evidenced by the summary in Appendix B. A parallel, complementary initiative spearheaded by Chuck Hopkins has resulted in the establishment of a number of ESD "Regional Centres of Expertise" (RCE's) in a number of jurisdictions. This initiative has its roots in one of the recommendations of chapter 36 of *Agenda 21*:

j) Countries, assisted by international organizations, non-governmental organizations and other sectors, could strengthen or establish national or *regional centres of excellence* [emphasis added] in interdisciplinary research and education in environmental and developmental sciences, law and the management of specific environmental problems. Such centres could be universities or existing networks in each country or region, promoting cooperative research and information sharing and dissemination. At the global level these functions should be performed by appropriate institutions.

RCE's have been organized in a number of countries under the auspices of the United Nations University. Each RCE is a network of existing formal, non-formal and informal education organizations, mobilized to deliver ESD to a regional community. Here in Canada, RCE's have been established in Saskatchewan, southern Ontario, and the Montreal region. Each RCE has a strong focus on research and an affiliation with one or more local universities. Environment Canada has provided support for the development of the Canadian RCE's.

Funding of ESD Initiatives

In several UNECE countries, UNDESD activities have received substantial financial support from the national government. Not so in Canada. The division of powers does complicate the ESD situation in Canada, but ESD extends beyond formal education, and federal funding of the strategic initiatives agreed to by the three partners (Manitoba, LSF, and EC) would not offend provincial sensibilities. Moreover many federal departments have as part of their core mandate "knowledge transfer" to the Canadian public (i.e. informal ESD). Furthermore, each federal department and key agency also has a Sustainable Development Strategy, one element of which typically involves public education; and each department is already (or should be in order to fulfill their SD commitments) engaged in non-formal SD education in the form of professional development and training of departmental employees.

Despite the clear link between ESD and the core mandate of nearly every department, the federal government has struggled to come up with the funds to support its international commitments to UNDESD. Soon after the UNDESD was announced, EC was given responsibility for this area by DFAIT. EC then entered the partnership with Manitoba and LSF outlined above, using the mechanism of a grants and contributions agreement with LSF to provide some funding for the agreed upon initiatives. But the total for each of the first two years was $325k, an amount that was subsequently cut nearly in half for the 2007. A commitment at this reduced level of $200k was forthcoming for 2008, with the indication that this would be the last subvention in support of the Decade (which of course continues until 2014). Spread across the entire country this is a pathetically small amount of support. Support for other ESD strategic initiatives was withdrawn so that the funds could be focused on NESDEC and the working groups. One unfortunate consequence of the funding cutback and the withdrawal of support for the other strategic initiatives (such as the ESD resource data bank and the sustainability curriculum review) was that LSF has had to seek funding from corporate partners[23] and even dip into its reserve funds to allow this work to continue. Thus a small NGO has shouldered the financial burden of funding some of Canada's international ESD commitments because the federal government was unable or unwilling to do so.[24] Another unfortunate consequence of the funding cutback was that even the modest sum of $15k per working group had to be cut back to $10k in 2008–09 in order to add funding for new working groups in PEI and Newfoundland and Labrador.

Provinces have varied considerably in their support for UNDESD. Although a supportive letter from the deputy minister of Education was one of the conditions of eligibility for potential provincial/territorial working groups (and therefore in each case they enjoyed at least that level of recognition/support from the provincial education ministry), some working groups have been hard pressed to garner additional Ministry or provincial government support. Where working groups have been most successful in getting provincial funding, it appears that they have enjoyed strong support from key politicians and a supportive senior management in the bureaucracy. (Manitoba is leading the way among anglophone provinces, with exciting recent developments in Ontario, Saskatchewan and British Columbia.)

ESD and Early Childhood Education

For many years those concerned with formal ESD have discussed the importance of starting the process as early as kindergarten. (LSF's original mandate, for example, referred to *K to 12*.) A conference entitled, "The Role of Early Childhood Education for a Sustainable Society" (jointly organized in Göteborg, Sweden, by Göteborg University, Chalmers University of Technology, and the City of Göteborg, from 2 to 4 May 2007, and featuring Chuck Hopkins as a keynote speaker) resulted in an important UNESCO publication which reported that:

There was a strong consensus that educating for sustainability should begin very early in life. It is in the early childhood period that children develop their basic values, attitudes, skills, behaviours and habits, which may be long lasting. Studies have shown that racial stereotypes are learned early and that young children are able to pick up cultural messages about wealth and inequality. As early childhood education is about laying a sound intellectual, psychological, emotional, social and physical foundation for development and lifelong learning, it has an enormous potential in fostering values, attitudes, skills and behaviours that support sustainable development – e.g. wise use of resources, cultural diversity, gender equality and democracy.[25]

ESD: Cure or Placebo?

Posing this dichotomy obviously begs the question about the existence of some sort of social illness or malady that requires a cure. Is planet earth in that much trouble? One is reminded of the cartoon that shows a doctor examining the planet and offering her diagnosis: "You've got humans!" Our high consumption lifestyle is destroying the ecosystems on which we (and many other species) depend. In a brilliant article discussing what he calls the "geopolitics of sustainability" economist Jeffrey Sachs points out that as a result of an 18–fold growth in the gross world product (GWP) over the past century:

Every major ecosystem, whether marine or terrestrial, is under stress. The world economy is depleting the earth's biodiversity, ocean fisheries, grasslands, tropical forests, and oil and gas reserves. We are massively and quickly changing the climate. These trends are occurring on a planet of 6.5 billion people and with economic activities that are already unsustainable as practiced. Yet with the economic successes now propelling India and China and the momentum of global population growth, we are on a trajectory to some nine billion people and a GWP of perhaps $275 trillion by mid-century.[26]

Not everyone agrees either that the predicament is as profound as this suggests; or that "humans" are to blame. (The sceptics cover a broad range of opinion, from those who claim scientific credentials [viz Bjorn Lomborg] to those who see our current problems as the unfolding of God's will.[27]) Nor should we be too hasty to predict the demise of planet earth. As the comedian George Carlin allegedly remarked, "The planet has been around for billions of years. It isn't going away. We are."

A more thoughtful objection might reject the medical analogy of illness/ disease and cure: too anthropocentric; too burdened by the ideology of medicalization. Allowing for all of these objections, is ESD capable of helping humankind "bend the curve" toward a more sustainable future, or does it – because it has a focus that is mid- or long-term rather than immediate – merely distract attention from the real challenges, in effect buying time for business as usual? I do not think there is an easy or straightforward response to this challenge. In some cases education has undoubtedly been put forward as a response to sustainability challenges that at the very least require more immediate action-oriented (perhaps more controversial, costly, or politically difficult) responses. This may make it appear to be a policy cop-out, a trivialization of the problem; or to go back to our medical analogy, a placebo.

While not suggesting that the opposite is true, that ESD is some sort of magic bullet that will miraculously "cure" the earth of the many sustainability issues we all face, I do believe strongly that insofar as it contributes to a sustainability "culture shift" ESD is an absolutely essential element of SD. It is a necessary, though not sufficient determinant of our success in finding a way to live more sustainably on this planet.[28] Not all sustainability enthusiasts agree. Some argue that we simply do not have time to wait for culture to shift[29] – we need action now, and this will require a more draconian approach involving regulations and other policy measures. A slightly different objection rests on the belief that all the requisite values already exist, but are latent. We do not need to change values, merely allow them to surface. A third variant is the argument that sustainability must be made to work with the existing culture by designing policies that can leverage sustainable behaviour through clever design of incentives and disincentives.

While each of these perspectives has merit, I reject the notion that either we educate or we use regulation and economic policy instruments. We need both. Education is not sufficient, but it is certainly necessary if we are to move toward a more sustainable future. At the same time, we must always remember that education is not the same as learning, which sometimes occurs despite educational efforts to the contrary. Properly purposed, education exists in the service of learning. The ultimate goal of ESD is to help "humans" learn to live more sustainably on this planet. In pursuit of this ambitious aim, ESD must effectively connect to the rapidly increasing power of the internet, which is quickly becoming the dominant mode of continuous learning in this, the age of "exponential times."[30]

CONCLUSION: WHERE DO WE NEED TO GO WITH ESD?

"If you are planning ahead 1 year, plant a seed.
If you are planning ahead 10 years, plant a tree.
If you are planning ahead 100 years, educate the people."

Confucian saying, c. 500 AD

1 I believe that ESD is basic, general education for the 21st century. Every citizen, every consumer, every employee, every decision maker,[31] every householder, needs to better understand the sustainability challenges and opportunities that will determine our fate. We should begin to measure "sustainability literacy" with at least as much vigour as we currently devote to other kinds of student performance.

2   Those who (in Max Weber's phrase) "have their hands on the wheel of history," who play important roles in organizations that shape our fate, need a more sophisticated level of comprehension of the complex inter-relationships between ecosystems and social systems, including the economy and the political system. (Note comments below on professional education.)

3   To help achieve this increased level of awareness and understanding, all three types of ESD (formal, informal and non-formal) will be required. But awareness alone is not enough. As it has from the beginning, ESD must pay attention to values, perspectives, skills, and practices. It must connect the head to the heart to the hands. Ultimately the benefit of ESD must be measured in terms of changed behaviour.[32]

4   With respect to formal education, ESD should indeed be infused across the curriculum – and modeled in institutional practices (the "whole school" approach) – at all levels, from pre-kindergarten to post doctoral studies. This will require a comprehensive change in policies, practices, curriculum, resources, and teacher training. Some might dismiss this recommendation as unrealistic, impossible. At this point we should remind ourselves of Kenneth Boulding's "Existence Theorem: everything that exists is possible!" The good news is that prototypes of nearly all of these elements have been developed. To a large extent, what is needed is not invention but application, which requires both the enactment of supportive policies and the development of appropriate standards.[33]

5   To the extent that this prescription is followed at the post-secondary level it will have important implications for qualification and training in all of the professions and trades.[34] I think this is a good thing. Prototypes and precursors of the requisite change are already evident in business schools, accounting, engineering, law, education, architecture, design, planning, medicine, social work, building construction and maintenance, urban infrastructure, HVAC, agriculture, landscaping etc. However, the application of sustainability-based professional education and training is limited and not yet across the board. Within 20 years (hopefully) ESD will be mainstreamed in all these areas for both "pre-service" and "in-service" professional education and training.

6   Societies like Canada may be on the cusp of a significant culture shift, not unlike the change that has taken place in the last 25 years in views about smoking. There is some survey data (Appendix C) to support this contention. One of the key drivers of this shift in culture will be informal ESD, especially the internet and the media. The impact of Al Gore's film *An Inconvenient Truth* illustrates how profoundly public awareness can be affected by this kind of production.[35] Another important development would be the use of advertising in support of SD rather than to encourage unsustainable forms of consumption.[36]

Am I being ridiculously optimistic? Should I heed the Russian proverb that defines a pessimist as an "informed optimist"? (Perhaps, like Humphrey Bogart in Casa Blanca, "I have been misinformed" about the prospects for a more sustainable future.) I think not. I very much like the comment David Orr made at a recent lecture at York University: "Hope is a verb with its sleeves rolled up." We need to roll up our sleeves, and despair is a very poor motivator.

APPENDIX A: SD INITIATIVES IN QUEBEC

Sustainable development is not a new idea within the Quebec government. Since the 1980's, action has been taken to raise awareness about the economic, environmental, and social principles and activities of living consistent with the idea of a sustainable future. In November 2004, the Ministry of the Environment released the *Québec Sustainable Development Plan, Consultation Document.* Over 3,500 people attended public hearings held in 21 municipalities across Quebec to discuss this document. Overall, the government received over 580 position papers and heard from more than 800 people. Following the consultations, changes were made to the draft legislation.

In 2006, the *Sustainable Development Act* came into force in Quebec to introduce a new management framework for the public service so that, when exercising its power and responsibilities, it integrates the principles of sustainable development. The Ministry of Sustainable Development, Environment and Parks is responsible for this Act, including its understanding, implementation, and the development of indicators to measure progress. The Quebec Ministry of Education, Recreation and Sports was involved in the development of the "Sustainable Development Plan" and is a member of the Inter-ministerial Committee on Sustainable Development, which ensures that government policy and practice are consistent with the principles of sustainable development and that the initiatives of the various departments, including those of education, are complementary.

In its first Sustainable Development Strategy (2008–13) the Government of Quebec has identified 3 priority strategic directions which include as Direction 1 "Inform, make aware, educate, innovate." Citing the global importance of the UN Decade for ESD, the Strategy documents states: "We must work even harder to make people aware of the concept of sustainable development, its imperatives and the environmental, social and economic challenges associated with it so that each member of society can help achieve it."

(*A Collective Commitment: Government Sustainable Development Strategy 2008 – 2013* (December 2007). Available at:
http://www.mddep.gouv.qc.ca/developpement/strategie_gouvernementale/
strat_gouv_en.pdf.)

APPENDIX B: ACTIVITIES OF CANADIAN
PROVINCIAL AND TERRITORIAL ESD
WORKING GROUPS[37]

The development of formal education projects/programs
- BCWG [British Columbia Working Group] in collaboration with The Ministry of Education, and BC Hydro has developed *Conceptualizing Environmental Learning: An Interdisciplinary Guide for Teachers* and a curriculum map to assist teachers of all subjects and grades to integrate environmental concepts into teaching and learning.
- BCWG is developing The *Taking Stock* document to look at what is happening with regard to sustainability at universities and colleges across BC (www.walkingthetalk.bc.ca/node/468).
- MESDWG [Manitoba Education for Sustainable Development Working Group] is creating ESD-focused education resources intended for the grade 12 level with the support of the Minister of Education and Manitoba Education Citizenship and Youth.
- MESDWG is partnering with the St. James School Division and Green Manitoba to support the development and dissemination of an ESD Resource Kit that the school division has developed for use within their division. The intent is to pilot and develop the ESD Resource Kit for implementation across Manitoba School Divisions over the next two years.
- SESDWG [Saskatchewan ESD Working Group] has organized four youth forums, bringing together more than 300 students to learn about sustainability issues and engage in action projects, 2007 Forums will take place in Saskatoon and Ile la Cross. SESDWG is exploring an opportunity to partner with Action Research: Community Problem Solving (AR:CPS) in Quebec to have access to Youth Action Forum programming for grade K – 12 in French.
- SENSE [Nova Scotia ESD Working Group called Sustainability Education in Nova Scotia for Everyone] has delivered Natural Step and Footprint programs to schools and communities in NS.
- EASO provided input into the Provincial Environmental Education policy, curriculum review and implementation. Through input into the Working Group on Environmental Education, Chaired by Roberta Bondar, *Shaping our Schools, Shaping our Future* contained 32 recommendations to strengthen ESD in Ontario schools. EASO provided input into the Environmental Education Policy and EE standards. EASO coordinated input from 14 organizations into the Science and Technology curriculum review, and coordinated stakeholders from the health and environment sector to provide input into the Health and Phy Ed curriculum review process.
- EASO, Ontario Teachers' Federation, Ministry of Education, and Ministry of Natural Resources facilitated the development and delivery of summer

camps to provide Ontario teachers with developmental professional learning on biodiversity and other sustainability issues in support of the Bondar report.
- NBWG has developed a green schools web portal for teachers, a resources kit to help schools/colleges initiate composting and energy conservation projects, they are sharing resources with teachers, and looking into a provincial green school policy, and funding proposals.

Projects and programs in non-formal ESD
- ABWG developing a carbon neutral eco-village.
- Sask WG is acting as advisory board for the new Green Life TV series – documentary style series to facilitate real and lasting change that will engage the Saskatchewan community in environmental issues.
- SENSE launched the Ecological Footprint project and Schools Facilities Management Greening program in schools across Nova Scotia SENSE has delivered Natural Step to businesses and communities.

Dialogue and networking
- BCWG www.walkingthetalk.bc.ca – communicative online gathering place which has 295 members representing 50 communities across BC and 11 communities internationally.
- ABWG – [Alberta Working Group]hosted a three day envisioning process with 35 formal, non-formal and informal education stakeholders to develop a shared vision, form groups with common themes, and develop action plans.
- MESDWG hosted an International ESD Conference in Winnipeg on November 26 to 28th, 2008 in collaboration with the Science Teachers Association of Manitoba. SESDWG hosted a Symposium on April 19 & 20, 2007 entitled Toward a Sustainable Future.
- SESDWG is exploring an opportunity to partner with Action Research: Community Problem Solving (AR: CPS) in Quebec to have access to Youth Action Forum programming for grade K – 12 in French.
- Education for Sustainable Development Networking Forum held on September 21, 2006 at Downsview Park engaged 180 individuals from 37 groups.
- SENSE hosted the First Annual Sustainability Education Symposium and Public Forum on Sustainability Education in March 2008.
- Saskatchewan International RCE meeting took place May 25th – 27th with Severn Cullis-Suzuki, from the David Suzuki Foundation, as the keynote speaker.

Communication to the broader public
- MESDWG is working with Green Manitoba and the Manitoba Forestry Association to facilitate a series of Educating the Educator workshops that

addresses topics of interest to the ESD and EL community. Four workshops have been presented to the public since June 2007.

- SENSE has created a Sustainable Development Resource Directory.
- SESDWG is creating www.saskesd.ca, a website serving as information portal and networking tool to advance sustainability education in the province.
- NBWG has created a website (www.nben.ca/seanb) and listserv for the working group.

ESD research

- MESDWG has in partnership with the International Institute for Sustainable Development, Manitoba Education Citizenship and Youth is undertaking research to expand the IISD Policy Bank Initiative as well as baseline data project to measure ESD attitudes pre and post decade.
- EASO, the Toronto Regional Centre for Expertise, the University of Toronto and Learning for a Sustainable Future have collaborated on the creation of a survey tool to collect data on ESD programs and activities through the NESDEC, WGs and RCEs.
- EASO Biodiversity Survey – During the fall of 2007, the Biodiversity Education and Awareness Network (BEAN) subcommittee of EASO surveyed providers of biodiversity education and awareness programs, materials and activities. One hundred fifteen groups responded – 30% of all responses were formal activities, 29% were non-formal activities and 41% were informal activities.

Modeling SD

- SENSE has launched the Ecological Footprint project and Schools Facilities Management Greening program in schools across Nova Scotia, and has delivered Natural Step to businesses and communities.
- SENSE has launched the Atlantic Canada Sustainability Initiative that is assisting organizations, municipalities and businesses with sustainability planning.

APPENDIX C: WHAT DO CANADIANS THINK OF SUSTAINABILITY?

http://www.ekostv.com/node/307

A survey released at Globe 2006 reveals that 53% of Canadians have never heard of the term "sustainability." Seven out of ten are unable to define the term sustainability. However, once the term is defined over 80% rated sustainability as a top or high priority national goal.

What we agree on

Survey highlights:
— 92 per cent of Canadians agree Canada should phase in mandatory standards requiring all new buildings and appliances to deliver 50 per cent more energy efficiency in 10 years.
— 89 per cent approve of meeting 100 per cent of Canada's new electricity needs through conservation measures, or renewable clean energy.
— 84 per cent agree that we need stricter laws and regulations to protect the environment.
— 83 per cent agree Canada should reduce taxes on income, payroll and investment, and replace these with taxes on pollution and depletion of natural resources.
— 83 per cent want the government to set strict national sustainability targets and report back to Canadians regularly on progress.
— 82 per cent agree Canada should introduce laws to promote denser, walkable cities that would make public transit more practical and reduce traffic congestion. Some 71 per cent want the same laws to protect farmland and reduce the environmental impacts of urban sprawl.
— 79 per cent approve of tax rebates on fuel-efficient vehicles funded by double GST paid on gas guzzlers not used for commercial or industrial purposes.
— 67 per cent agree that Canadians consume more than our share of world resources.
— 64 per cent disagree that protecting the environment usually means sacrificing comfort and convenience.

These are just some of the results of the public opinion poll conducted by James Hoggan and Associates for BC Hydro, Alcan, David Suzuki Foundation, Ethical Funds and several other organizations. The provocative findings were made public for the first time at Globe 2006 on March 31, 2006 in Vancouver B.C.

To access the full report go to:
James Hoggan and Associates www.hoggan.com

Vancouver Sun, 31 March 2006
Section: BC Business
By M. Kane, Page H3

People "think they're alone" in wanting to save the planet: Eight out of 10 Canadians want tougher laws to protect environment. Canadians care strongly about saving the planet but wrongly believe that many of their fellow citizens do not, according to survey findings to be released today. They blame inadequate information and a lack of government leadership for their own failure to behave sustainably while assuming that others are not really concerned, said James Hoggan, a public relations expert. "They think, 'Well, why should I be the chump who behaves in a more environmentally friendly manner when no one elsc is?' But that's because they have this mistaken view that people outside their circle of friends really don't care. That's actually not true."

The survey shows that more than eight in 10 Canadians believe the government should enact stricter laws and regulations to support a more sustainable economy that protects and manages the country's resources for future generations. They also want taxes shifted to those who pollute and deplete natural resources, and double GST slapped on gas guzzlers to fund tax rebates for fuel-efficient vehicles. The findings, to be presented today at the Globe 2006 environmental business fair in Vancouver, were characterized by Hoggan as both a wake-up call and a leadership opportunity for government and business. He said advocates of sustainable business practices "are guilty of talking to themselves a lot" instead of reaching out to the public who drive public policy and consumer patterns. "One of the key problems that we have in Canada is that Canadians do one thing and say another," Hoggan said in an interview. "Anybody who is trying to advocate sustainable behaviour should try to weave in the message that Canadians actually do care."

He said the survey demonstrates that the great majority of Canadians want the economy to be successful today while sustaining the country's environmental, economic and social resources for future generations. Asked why Canadians do not behave more sustainably, 48 per cent blame government leadership first. More than seven in 10 Canadians surveyed agree with the statement: "If everyone in the world lived the consumer lifestyle we enjoy in North America, we would destroy the planet." Hoggan also said companies and organizations need to do a better job of explaining the term "sustainability," which is defined as development that meets the needs of the present without compromising the ability of future generations to meet their own needs.

The survey is part of the Sustainability Research Initiative led by James Hoggan & Associates, a public relations agency extensively involved in environmental and sustainability issues, and the Globe Foundation, which promotes the

$1.1 trillion international business of the environment. Sponsors include Canadian Pacific Railway, BC Hydro, the University of B.C., the David Suzuki Foundation, Vancouver-based Ethical Funds, Greater Vancouver Regional District, Concord Pacific, the Fraser Basin Council, Alcan and the Port of Vancouver. The sustainability survey of 2,500 Canadians was conducted by McAllister Opinion Research of Vancouver in all provinces except Quebec. The findings are considered accurate within two percentage points, 19 times out of 20.

NOTES

1 I am grateful to Christina McDonald for comments on an earlier draft of this paper.
2 World Commission on Environment and Development (WCED), *Our Common Future* (Oxford University Press, 1987), 113. The Report notes the importance of teacher training to provide this new kind of education, calling it a "critical point of intervention."
3 Ibid., xiv.
4 The 10 individuals chosen to prepare a draft of chapter 36 included Canadian Chuck Hopkins, who at the time was a superintendent for Curriculum with the Toronto School Board. After retiring from this position Chuck has continued to serve as a key senior ESD advisor to UNESCO, and in 1999 was appointed UNESCO/UNITWIN chair in "Reorienting Teacher Education Toward Sustainability" at York University.
5 One very noticeable change since 1992 is the move away from the rather awkward language of "environment and development" to the more comprehensive conceptualization of ESD as involving integrated, systems-based understanding of the interrelationships between the natural, social, and economic dimensions of sustainability. The seeds for this broadening were present in chapter 36 however, which spoke of connecting environment and development issues to "their socio-cultural and demographic aspects and linkages."
6 See further discussion of this point below. Chuck Hopkins is particularly adamant about avoiding the term "sustainability education." He has compiled a list of various forms of "adjectival education" (environmental, global, driver, health etc) that are constantly vying for inclusion on the curriculum. Sustainability education would be added to this list of over 80 others, forced to wait at the end of the line of never. ESD is not another topic to be added to the growing, unmanageable agenda for curriculum expansion; on the contrary it is a different approach to the entire curriculum, both explicit and "hidden." (The latter term refers to the various institutional practices followed in school which should be reformed to ensure that the school itself "models" sustainability practices.)
7 Rosalyn McKeown (with assistance from Charles Hopkins et al), *The ESD Toolkit*, 24–25. Available at: www.esdtoolkit.org. This document provides an excellent overview of the origins and evolution of ESD, and contains useful advice about its application.
8 Ibid.

9 John Fien and Daniella Tilbury, "The Global Sustainability Challenge," in Daniella Tilbury et al., eds. *Education and Sustainability: Responding to the Global Challenge* (IUCN, 2002). Available at: www.unece.org/env/esd/information/Publications%20IUCN/education.pdf.

10 McKeown, *ESD Toolkit*, 7.

11 Of course ESD, like sustainability itself, has many precursors. In terms of UN antecedents of ESD none is more important than the conference on Environmental Education (co-sponsored by UNESCO and UNEP) held in Tblisi Georgia (USSR) in October 1977. Many of the elements of ESD are reflected in the Tblisi Declaration which emanated from the conference. Available at: www.gdrc.org/uem/ee/tbilisi.html.

12 It is no coincidence that Manitoba is now playing a lead role in Canada's response to the UN Decade of Education for Sustainable Development (UNDESD).

13 In her recent doctoral dissertation, Liza Ireland observes "Canadian environmental education scholars such as Jickling (1992, 2001) and Sauvé (1999) have interpreted education for sustainability in narrow, instrumental terms and have criticized it, as 'sustainable development' is a contested term and education 'for' sustainable development can be interpreted as indoctrination. In light of this they have not adopted the new terminology of 'education for sustainability' or 'education for sustainable development' but have retained the term 'environmental education' seeing it as more appropriate." Liza Ireland, "Educating for the 21st Century: Advancing an Ecologically Sustainable Society," PhD Dissertation (University of Stirling, November 2006). Elements of this controversy continue to flare up in print as well as at meetings and conferences. A full consensus among educators to support ESD may eventually emerge. In my view the controversy has not been helpful. ESD has been an evolving, deliberately inclusive concept which must in any event be adapted to different regions and changing SD challenges and opportunities.

14 See Learning for a Sustainable Future. Available at: www.lsf-lst.ca.

15 The group within IUCN that oversees its involvement in ESD is called the Commission on Education and Communication (CEC). IUCN's website contains the following statements about ESD. "The Commission is taking part in a global effort to integrate the principles of sustainable development into all aspects of education and learning. It's called Education for Sustainable Development (ESD). ESD focuses on how people live, work and make decisions. All ESD efforts share a common concern for a sustainable future, employing a wide variety of methods that engage people in problem-solving and advance positive change." CEC's website is available at: http://cec.wcln.org/index.php?module=pagesetter&func=viewpub&tid=11&pid=124; accessed 7 October 2007.

16 Thessaloniki Declaration, paragraph 10. Available at: www.mio-ecsde.org/old/Thess/declar_en.htm. As Fien and Tilbury argue: "Education with the objective of achieving sustainability varies from previous approaches to environmental education in that it focuses sharply on developing closer links among environmental quality, human equality, human rights and peace and their underlying political threads. Issues such as

food security, poverty, sustainable tourism, urban quality, women, fair trade, green consumerism, ecological public health and waste management as well as those of climatic change, deforestation, land degradation, desertification, depletion of natural resources and loss of biodiversity are primary concerns for both environmental and development education. Matters of environmental quality and human development are central to education for sustainability." Fien and Tilbury, "The Global Sustainability Challenge," 9.

17 *"Educating and Acting for a Sustainable Future"* (nd) published by the Centrale des syndicats du Quebec (CSQ). The quote in the passage is taken from André Beauchamp, "ENVIRO-SAGE, Vers un politique d'information, de sensibilisation et d'éducation à la gestion durable des matières résiduelles. Document de réflexion soumis à la société," RECYC-*Québec* (January 1999), 21.

18 As stated on the UNESCO website. Available at: http://portal.unesco.org/education/admin/ev.php?URL_ID=27234&URL_DO=DO_TOPIC&URL_SECTION=201.

19 DFAIT had responsibility initially for UNDESD, but passed it on to Environment Canada within a few months.

20 Although it participated in this key meeting, Quebec has so far declined an invitation to form a provincial working group and join the National Education for Sustainable Development Expert Council (NESDEC). Nevertheless as pointed out above Quebec has a number of strong ESD activities and programs in place, and did report on some of these activities by providing data to Council of Ministers of Education of Canada (CMEC) for inclusion in its report to UNECE.

21 The first CMEC report was released in October 2007 under the title "Report to UNECE and UNESCO on Indicators of Education for Sustainable Development." In addition to reporting on progress toward achieving the objectives note above, countries were also asked to "Describe any challenges and obstacles encountered in the implementation of the strategy" and to "Describe any assistance needed to improve implementation." Absent the activities around UNDESD, it is unlikely that CMEC would have had much if any formal involvement in ESD. In 1999, Manitoba tried unsuccessfully to encourage CMEC to develop an action plan for implementation of ESD in each jurisdiction. In June 2008, however, CMEC did agree to establish a Working Group on ESD that will be led by Manitoba.

22 Information in the Appendix A was drawn in part from the CMEC Report.

23 Hewlett-Packard (HP) has helped fund the R4R initiative and Cadbury Schweppes has provided support for a new round of Youth Taking Action Forums aimed at senior public/junior high students.

24 Several EC officials worked long and hard to win support from their department for ESD among competing priorities in a period of internal turbulence. Their efforts are deeply appreciated in the ESD community. Although the Harper government in Canada has provided meager support for ESD, ironically the Bush government in the US recently approved a massive infusion of funds by signing The Higher Education Sustainability Act (HESA), approved by Congress as part of the new Higher Education Opportunity Act of 2008 (HR 4137) which created a pioneering

"University Sustainability Grants Program" at the Department of Education. This program will offer competitive grants to higher education institutions and associations to develop and implement sustainability curricula, practices, and academic programs. (See the announcement at: http://ncseonline.org/Updates/cms.cfm?id=2458)

25 *The Contribution of Early Childhood Education to a Sustainable Society.* Ingrid Pramling Samuelsson and Yoshie Kaga eds. (UNESCO, 2008). Available at: http://unesdoc.unesco.org/images/0015/001593/159355E.pdf.

26 Jeffrey Sachs, "The New Geopolitics." *Scientific American* (June 2006). Available at: www.sciam.com/article.cfm?chanID=sa006&articleID=000B4E50–A56D-146C-A56D83414B7F0000&colID=31. Sachs goes on to comment that "Our global politics is not yet adapted to the challenges of sustainability." To do so we must develop new approaches to governance and global politics "based firmly on the budding science of sustainability."

27 The latter group would likely comprise many of those who (like vice-presidential nominee Sarah Palin) believe in Creationism. In a recent Gallup poll done in the US, when presented with 3 contrasting statements about the origin of the universe 45% of respondents chose the statement that most closely describes biblical creationism: "God created human beings pretty much in their present form at one time within the last 10,000 years or so." A slightly larger percentage, almost half, chose one of the two evolution-oriented statements: 37% selected "Human beings have developed over millions of years from less advanced forms of life, but God guided this process" and 12% chose "Human beings have developed over millions of years from less advanced forms of life, but God had no part in this process." The American public has not notably changed its opinion on this question since Gallup started asking it in 1982. See www.unl.edu/rhames/courses/current/creation/evol-poll.htm; accessed 14 Oct. 2007.

28 Here again I may be accused of begging the question as to whether there is any truly sustainable form of development; or whether SD holds out any hope of helping usher in a more sustainable future for humankind.

29 It would be a serious error to ignore the profound shifts in culture and social structure that are already underway in all parts of the globe, driven to a large extent by economic globalization and the transformation of electronic communications, See for example the wildly popular You Tube video "Shift Happens." Available at: www.youtube.com/watch?v=pMcfrLYDm2U.

30 This phrase appears in the video "Shift Happens" noted above.

31 Of course we are all decision makers in our everyday lives. We make hundreds if not thousands of decisions every day from the moment we wake up until we fall back asleep. We decide where to live, where to work, how to get there, what clothes to wear, what food to eat, what forms of entertainment to enjoy, whether to vote and for whom, etc. These myriad decisions ultimately drive the economy, the political system, and other social systems.

32 IISD and the Manitoba government are currently undertaking a longitudinal survey to try to assess the impact on behaviour of various forms of ESD.

33 Ontario has recently developed new "standards" for environmental education as a follow-up to the recommendations of the Bondar Working Group.

34 Note the following quotation from the Foreword to a recent report on sustainability initiatives in us colleges and universities: "The men and women who, in 20 years, will lead our businesses, educational institutions and government agencies are in school now. We need to offer them the kind of academic and professional preparation that will ready them to envision and create a different kind of world. It will be a world which has new and cleaner forms of energy production, transportation, agriculture, natural resource management, health care, scientific research, micro and macro businesses, and other essential technological advances. To achieve this at the speed required will call for serious new support including new guidance and funding from federal and state governments, and a complete rethinking of how we educate every degree candidate from architecture and engineering to accounting and even teaching itself." *Campus Environment 2008: A National Report Card on Sustainability in Higher Education. Trends and New Developments in College and University Leadership, Academics and Operations* Second edition (August 2008). Available at: www.nwf.org:80/campusEcology/campusreportcard.cfm.

35 A controversy arose in the uk regarding the government's plan to distribute Al Gore's film to classrooms for required viewing. A court challenge resulted in a verdict that allowed the plan to go ahead with the proviso that some of the inaccuracies and exaggerations in the film be noted and discussed.

36 Chuck Hopkins brokered a "Type 2 Partnership" in support of esd at the World Summit on Sustainable Development with advertising conglomerate J. Walter Thompson, whose ceo acknowledged that hundreds of billions of dollars have been spent (particularly in the us) to promote unsustainability. For various reasons the partnership did not come to fruition, but future agreements may be more successful.

37 This information is taken from the "Proposal to Environment Canada From Learning for a Sustainable Future (lsf) in support of the National Education for Sustainable Development Expert Council (nesdec) and Provincial/Territorial Education for Sustainable Development Working Groups (March 2008)." By 2008 working groups had been established in Alberta, Nova Scotia, New Brunswick, Ontario, Manitoba, Saskatchewan, British Columbia, pei, and Newfoundland and Labrador.

8 Canadian Business and the Sustainability Challenge: Engagement and Performance

DAVID WHEELER AND
ANNIKA TAMLYN

By now even the most hardened sceptic admits it: Global warming and other environmental and health challenges are going to require fundamental changes in the way we live and work as a society. It's no longer a matter of choice or opinion, but of survival.

In a consumer-based economy like ours, that means no sector has more at stake than retail. Think of it. The retail sector is Canada's largest employer. We're the largest consumer of energy. We produce more waste than any other sector. Our physical store layouts and their geographic locations, our supply and distribution networks, and of course, the products we carry all have a huge environmental impact.

Galen Weston Jr, Globe and Mail, 23 April 2007

Business and industry have a crucial role to play in achieving sustainable development in Canada.[1] If we examine just one issue – climate change – it is salient to note that Natural Resources Canada (NRCan) estimated that in 2003 Canadian industry accounted for 38% of energy usage in Canada and 34% of greenhouse gas (GHG) emissions.[2][3] Environment Canada's 2005 report on GHG emissions[4] summarized data based on industry sectors from 1990 to 2005. In that period there was an overall 25.3 percent *increase* in emissions. The greatest increase associated with stationary sources of energy consumption was in the mining sector (151.9%), followed by the fossil fuel industries (42.5%) and commercial and institutional buildings (42.5%). Electricity and heat generation emissions increased by 34.9 percent but they accounted for the highest volume from all stationary sources at 37 percent of the total. These are very challenging statistics for a country that is a signatory to the Kyoto protocol, and they do emphasise the vital and urgent importance of engaging business in addressing Canada's sustainability as a country. From a governance perspective, there is a fundamental dilemma as to exactly who in business policy-makers should engage with on the broad range of sustainability issues: climate change, toxic emissions, biodiversity, economic development, fair trade, human capital development, social inclusion, poverty, aboriginal rights and a host of other pressing questions. Unfortunately, the answer to this question in neither simple nor static.

In recent years a number of Canada's best known sustainable business icons saw their ownership transfer out of the country. On February 21st, 2006 Arcelor announced a successful takeover of Dofasco having obtained 88.4% of outstanding shares. On March 14th, 2006 General Electric agreed to acquire ZENON Environmental for CAD $760 million. On 21 August 2006 Xstrata announced a full takeover of management of Falconbridge. Falconbridge had previously merged with Noranda on 30 June 2005. On 5 January 2007 shares of Inco Ltd stopped trading on the Toronto Stock Exchange after Companhia Vale do Rio completed its takeover of the company. On 25 April 2007 Shell Canada delisted from the Toronto Stock Exchange following a successful buy-back by parent Royal Dutch Shell plc. And on 28 September 2007, Rio Tinto shareholders backed a takeover of Alcan by 708.7 million votes to 19.6 million.

The removal of Dofasco, ZENON, Falconbridge/Noranda, Inco, Shell Canada, and Alcan from their Canadian owners in the space of less than two years signalled a significant shift in governance and power relations around a number of high visibility companies in the sustainable development discourse in Canada. Whether this is a net positive or net negative effect for Canada and the rest of the world from a sustainability perspective is one of the questions we will address in this paper. We will also examine implications for the sustainability of Canadian business from the perspective of less visible and less "high impact" firms and sectors, including small and medium sized enterprises. But let us start by examining what motivates business to adopt sustainable practices in the first place.

WHY DO BUSINESS ORGANIZATIONS EMBRACE SUSTAINABILITY?

Kurucz and co-workers[5] have provided a typology of "business cases" for corporate social and environmental responsibility, each involving a distinct value proposition: i) cost and risk reduction; ii) profit maximization and competitive advantage; iii) reputation and legitimacy; and iv) synergistic value creation. Consistent with the arguments of Aragón-Correa and Sharma[6] and Rubio-Lópes[7] who have drawn attention to some important "myths and misunderstandings" in corporate environmental strategizing, Kurucz et al explain why there can be no generalizable business case for sustainability. In contrast to earlier exhortations based on strategic, managerial, or moral reasoning[8] and more recent popular calls to business action,[9] Kurucz and co-workers propose a more nuanced understanding of motivations which places business organizations more firmly in their twin roles of social actor and value creator and emphasise the need for pragmatic, experimental action at both organizational and societal levels.

Some of the more interesting research on the phenomenon of corporate greening has been conducted by Canadian researchers, in particular with respect to "high impact" sectors eg oil and gas,[10] electricity generation,[11] and forest products.[12] Organizational context, strategic imperatives, managerial perceptions, stakeholder influences, and corporate values are all cited as potentially important factors in the adoption of sustainability practices in these sectors. In a meta-exploration of Canadian forestry, oil and gas and mining companies" motivations between 1986 and 1995, Tima Bansal[13] found that two of the most salient reasons for corporate greening in these sectors were resource-based (i.e. an internal strategic logic) and institutional (e.g. the role of external influencers such as the media and within-industry mimicry).

Beyond Canada, institutional theorist Andrew Hoffman[14] has described how external forces impacting the us chemical industry shaped that industry's responses to the rise of environmental concerns in the last decades of the 20th century. Hoffman drew attention to the importance of "disruptive events" in re-shaping the institutional field and the growing importance of external actors, including regulators and non-governmental organizations, in defining the subject of discourse and power relations within a sectoral field. Similarly, and echoing to an extent stakeholder theory and social capital arguments,[15] Aguilera et al[16] have proposed a multi-level model for explaining why firms adopt corporate social responsibility practices with respect to individual, organizational, national, and transnational actors" instrumental, relational, and moral interests. Aguilera and colleagues argue, among other things, for the importance of legitimation as a driver of firm responses and the influence of external actors e.g. governments and non-governmental organizations in driving up standards for entire sectors e.g. the food industry in Europe and the chemical and extractive industries internationally.

So there appears to be a growing appreciation in the academic literature of the complexity of corporate greening phenomena; and happily we are beginning to move beyond the "one size fits all" exhortations. Of course, in Canada, as in several other jurisdictions, it is becoming very difficult for "high impact" corporations to ignore their sectoral and societal context and the growing expectations of external actors for social and environmental responsibility. It must be noted however, that worldwide we do not have anything like the level of understanding of institutional greening phenomena in less visible corporations and sectors, especially those with lower sustainability impacts.[17] And with very few exceptions[18] we are almost totally lacking in equivalent insights for the motivation of small and medium sized firms in Canada, except perhaps for their responses to supply chain or regulatory compliance demands.

Given the systemic bias towards studies of large companies in high impact sectors, it is no simple matter to assess the past, present, and potential future contributions of Canadian business to sustainable development domestically and internationally. Thus, in order to address the topic we adopt a "twin track" approach in this paper. First we examine the jurisdictional (federal) level to assess the degree to which corporate Canada is i) engaged; and ii) performing in a manner consistent with principles of sustainable development. Second, we examine the case for adopting special policy instruments for small and medium sized enterprises based on their collective sustainability impacts and what little we do know of their responsiveness to internal and external sustainability drivers.

CORPORATE CANADA AND SUSTAINABILITY ENGAGEMENT AND PERFORMANCE

The degree to which different jurisdictional and cultural norms affect corporate attitudes to social, environmental and ethical business practices is beginning to be recognised, if not wholly understood.[19] In order to provide a basis for comparison between Canadian firms and their international counterparts in sustainability issues we developed an index based on five sub-indices of corporate sustainability *engagement* and corporate sustainability *performance*. The rationale for including sustainability engagement as well as performance was that engagement is a surrogate indicator for responsiveness to external institutional forces as well as a signal of commitment of organizational resources to strategic learning, communication and social capital (goodwill) building activities, all of which demand leadership-level involvement.[20] The sub-indices were selected on the basis of four criteria: i) direct or indirect linkage with corporate sustainability engagement; ii) direct or indirect linkage with corporate sustainability performance; iii) transparency and ease of data acquisition; and iv) potential for periodic updating for purposes of future tracking. In each case the sub-indices were normalised against size of national GDP for the previous year. Ten countries were included in the overall index: the G8 countries i.e. Canada, France, Germany, Italy, Japan, the Russian Federation, the United Kingdom, and the United States, plus two extra European countries: the Netherlands and Sweden, because of their national reputations for business and societal enlightenment on matters of sustainable development.[21] The sub-indices were:

1) membership of the World Business Council for Sustainable Development (WBCSD);
2) active membership of the United Nations Global Compact;
3) total registrations to environmental management systems standard ISO 14001;
4) total cumulative publications of Global Reporting Initiative compliant sustainability reports; and
5) membership of the Dow Jones Group Sustainability (World) Index.

All of these sub-indices meet criteria (iii) and (iv) – transparency and trackability. And so bearing in mind criteria (i) and (ii) – direct or indirect linkage with corporate sustainability engagement or performance – the reasoning for the inclusion of each of the sub-indices was as follows:

Membership of the World Business Council for Sustainable Development (WBCSD)

As of September 2007, the WBCSD had in excess two hundred member organizations around the world and is arguably the pre-eminent corporate membership organization representing sustainable business globally. Formed during the middle years of the 1990s from a merger of the Business Council on Sustainable Development and the World Industry Council on the Environment, WBCSD played a vital catalytic role – together with the more conservative International Chamber of Commerce (ICC) – in mobilising the business voice at the 2002 World Summit on Sustainable Development (WSSD) in Johannesburg. According to the 2006 Globescan "Survey of Experts" the WBCSD web site was viewed as "by far" the best source of information on sustainable development, ahead of civil society organizations such as the International Institute for Sustainable Development and the World Resources Institute.[22] For these and a number of other reasons[23] we believe this sub-indice is highly and directly relevant to the corporate engagement criterion and indirectly related to the performance criterion. Perceived sustainability laggards are most unlikely to be invited to join WBCSD.

Active Membership of the United Nations Global Compact

The United Nations Global Compact was launched following a speech by Secretary General Kofi Annan at the World Economic Forum in 1999.[24] As of September 2007, more than 3000 organizations were "active members" globally. An active member is a member that not only embraces the ten principles of the Compact, but which is also able to provide a report on progress against the principles within three years of joining. We believe that this sub-indice is of direct relevance from a corporate engagement perspective in that it requires a corporate level commitment to a set of international standards with growing credibility. The criterion is perhaps less relevant to the performance criterion, although in due course it is possible that the reporting requirement will drive performance in an indirect way.

Total Registrations to Environmental Management Systems Standard ISO 14001

The International Organisation for Standardization (ISO) developed a series of environmental management systems standards in response to increased corporate interest in environmental issues in the 1990s. Both non-governmental and regulatory agencies have drawn attention to the danger that management systems standards provide no guarantee of legal compliance, still less high levels of environmental performance.[25] And yet the growth in the general management systems standard (ISO 14001), to nearly 130,000 worldwide registrations by September 2007, must indicate some level of deepening acknowledgement of the value of minimizing environmental impacts within certain sectors. Thus we believe this sub-indice has direct relevance from a corporate engagement perspective in that it reflects the need to signal to supply chain partners, corporate parents and sectoral opinion-formers that environmental issues are being addressed. In this way there may be a reduction in environmental risk or reputational risk to the firm's clients and head offices and therefore a potential indirect benefit to performance.

Total Cumulative Publications of Global Reporting Initiative Compliant Sustainability Reports

The Global Reporting Initiative (GRI) was launched in 2002 based on an initiative of the Coalition for Environmentally Responsible Economies (CERES) and a number of other organizations keen to see greater standardization in environmental and social reporting.[26] In 2006 GRI launched version three (G3) of its guidelines for sustainability reporting and as of September 2007 just under two and one half thousand compliant reports had been produced and included in the GRI database. Today GRI is a broad-based multi-stakeholder institution, formally affiliated with the United Nations Environment Program, and it may be argued that the G3 guidelines for reporting have become the de facto standard for sustainability reporting globally. We suggest that publication of a GRI report has significant direct relevance from a corporate engagement perspective; principally because production of such a report requires a broad range of activities in data collection, validation and reporting across the organization and the inevitable direct involvement of senior management. Any linkage to performance is likely to be indirect although of course symbolic acts like the release of a fully transparent, GRI compliant report should ideally re-enforce substantive acts related to performance in order for legitimacy to result.[27]

Membership of the Dow Jones Group Sustainability (World) Index

The Dow Jones Group Sustainability Indices (DJGSIs) were launched in 1999 and since that time have become some of the best known indices tracking the performance of sustainable businesses worldwide. The indices are a joint venture between Dow Jones Indices, STOXX Ltd of Europe and Sustainable Asset Management, a Zurich-based ratings and research organization. In 2007 there was approximately us$5 billion under management through sixty licensees of the indices around the world, of which three were Canadian: Credit Union Central Alberta Ltd, Credit Union Central of Ontario, and TD Asset Management. To enter an index, a company must feature in the top 10 percentile of performers in their sector on the basis of a range of policy-related and direct performance. The World index is updated annually and in September 2007 included a total of 309 companies. There is mixed evidence that membership of the DJGSI generates a priori financial or other competitive advantage benefits[28] or that the largely self-reported nature of data used to assess the inclusion of firms in the index provides a reliable and direct representation of performance.[29] Nevertheless, we believe that this sub-indice provides indirect evidence of sustainability engagement because i) many of the criteria are based on policy and reporting factors that are unequivocally indicators of engagement; and ii) it is typical for corporations to celebrate their inclusion in the index as a mark of their sustainability credentials. We also believe that unlike most of the sub-indices used in our ranking, membership of the Dow Jones Group Sustainability (World) Index does offer some *direct evidence* of performance in a range of environmental, social and human resource practices, notwithstanding the self-reported nature of some of the metrics.

In summary, we have selected a range of indices, further described elsewhere,[30] which together provide both direct and indirect evidence of corporate sustainability, engagement, and performance. For our purposes here, they also all satisfy our criteria for transparency and ease of data acquisition and potential for periodic updating for purposes of future tracking. In that sense they may be judged stable and well recognised indices that provide a balanced picture of corporate sustainability performance and engagement. In order to establish our overall index we calculated the percentage of Canadian participation in each sub-indice. We then normalised this performance for size of GDP and then transformed the data so that 50% over-performance translated to a national score of 1.5, and 50% underperformance translated to a national score of 0.5. We then took the arithmetic and geometric means of the scores and produced a ranking. Only the geometric means are cited here.[31] Results are depicted below and demonstrate that in September 2007 corporate Canada fared better than the international norm in two sub-indices (GRI compliant reporting and membership of the DJGS (World) Index), and less well in the remaining sub-indices, with especially poor performance in the sub-indice on active membership of the UN Global Compact. Overall corporate Canada places sixth in the ranking (and 4th out of the G8) with a geometric mean performance factor of 0.825 compared with 2.266 for first placed Sweden and 0.185 for the 10th placed Russian Federation.

The table provides some insight into the position of corporate Canada compared with its counterparts in nine other jurisdictions in 2007. It is perhaps less than surprising that corporate Sweden and corporate Netherlands do so well;

Table 1
Evaluative Corporate Sustainability, Engagement, and Practice

Country	GDP[59] (% world total)	Membership of WBCSD[59] (% world total) Performance factor normalised by GDP	Active Membership of Global Compact[59] (% world total) Performance factor normalised by GDP	ISO 14001 Registrations[59] (% world total) Performance factor normalised by GDP	Global Reporting Initiative Compliant Reports[59] (% world total) Performance factor normalised by GDP	Dow Jones Group Sustainability (World) Index[59] (% world total) Performance factor normalised by GDP	Geometric Mean Performance Factor [and Rank]
Canada	1.251 (2.59)	5 (2.42) 0.934	22 (0.72) 0.278	2578 (2.00) 0.772	101 (4.11) 1.587	10[59] (3.24) 1.251	0.825 [6th]
France	2.231 (4.62)	9 (4.34) 0.939	306 (9.98) 2.160	3629 (2.81) 0.608	114 (4.64) 1.004	17 (5.50) 1.190	1.075 [4th]
Germany	2.907 (6.03)	12 (5.80) 0.962	88 (2.87) 0.476	5800 (4.50) 0.746	85 (3.46) 0.574	19 (6.15) 1.020	0.720 [8th]
Italy	1.845 (3.82)	6 (2.90) 0.759	74 (2.41) 0.631	9825 (7.61) 1.992	91 (3.70) 0.969	5 (1.62) 0.424	0.824 [6th]
Japan	4.340 (9.00)	29 (14.0) 1.556	53 (1.73) 0.192	21779 (16.88) 1.876	279 (11.36) 1.262	39 (12.62) 1.402	0.994 [5th]
Russian Federation	0.987 (2.05)	3 (1.45) 0.707	9 (0.29) 0.141	223 (0.17) 0.083	8 (0.33) 0.161	0 (0) 0	0.105 [10th]

Country	GDP[59] (% world total)	Membership of WBCSD[59] (% world total) Performance factor normalised by GDP	Active Membership of Global Compact[59] (% world total) Performance factor normalised by GDP	ISO 14001 Registrations[59] (% world total) Performance factor normalised by GDP	Global Reporting Initiative Compliant Reports[59] (% world total) Performance factor normalised by GDP	Dow Jones Group Sustainability (World) Index[59] (% world total) Performance factor normalised by GDP	Geometric Mean Performance Factor [and Rank]
United Kingdom	2.345 (4.86)	7 (3.38) 0.695	78 (2.54) 0.523	5400 (4.19) 0.862	209 (8.51) 1.751	67 (21.68) 4.498	1.194 [3rd]
United States of America	13.202 (27.36)	42 (20.29) 0.742	92 (3.00) 0.110	8081 (6.26) 0.229	244 (9.93) 0.363	57 (18.45) 0.674	0.337 [9th]
Netherlands	0.658 (1.36)	9 (4.35) 3.200	24 (0.78) 0.574	1132 (0.88) 0.647	118 (4.80) 3.529	14 (4.53) 3.331	1.687 [2nd]
Sweden	0.385 (0.798)	3 (1.45) 1.817	33 (1.08) 1.353	4865 (3.77) 4.724	50 (2.04) 2.556	5 (1.62) 2.030	2.266 [1st]
World	48.245 trillion	207	3067	129031	2456	309	

with significant out-performances *versus* GDP evident in all five sub-indices for Sweden and three out of five for the Netherlands. It is also perhaps unsurprising that the US and the Russian Federation score less well, with relatively poor engagement and performance in all five sub-indices, the US being significantly under-represented in active membership of the UN Global Compact, ISO 14001 registrations and GRI compliant reporting. Perhaps less intuitively, French corporations outperform their German counterparts overall, largely due to French companies featuring well on UN Global Compact membership and GRI compliant reporting criteria. French and German companies are very similarly represented in WBCSD membership and DJGSI inclusion criteria. Corporate Canada neatly bisects its French and German counterparts, scoring creditably in GRI compliant reporting and DJGSI inclusion.

In order to improve its ranking, a simple recommendation for corporate Canada arising from these data might be to increase engagement with the UN Global Compact and/or increase the number of ISO 14001 registrations. For example, raising the number of active corporate memberships of the UN Global Compact from a relatively low 22 to say, 100 (a number still only one third of the total for France), would move the overall geometric mean national performance to 1.12. Canada would then leapfrog to fourth place in the ranking, just behind the UK. But this is not really the point. For Canadian corporations to join the Global Compact in such numbers would denote a significant shift in current institutional perceptions and/or signalling requirements. It is likely that this would only occur if a serious reason emerged to change current perceptions and/or the need for signalling behaviours e.g. new leadership acts by CEOs and/or some national or sectoral initiative that related membership of the Compact to corporate advantage.

SMALL AND MEDIUM SIZED ENTERPRISES
AND SUSTAINABILITY ENGAGEMENT AND
PERFORMANCE[32]

To date, the business and sustainability literature has largely overlooked the performance of small businesses.[33] However, there is clearly a need to measure and evaluate the impacts of SMEs on the environment and on society, and to identify what motivates all businesses – including SMEs – to become "green."[34] Here we explore the proposition that the collective sustainability impacts of Canadian SMEs may be sufficiently important to the future sustainability challenge for Canada that special measures may be required to take advantage of the opportunity they represent.

SMEs *and the Economy*

SMEs contribute to a major portion of economic activity in most nations, especially in terms of employment.[35] According to Industry Canada's *Small*

Business Quarterly, as of June 2006 there were over 2.3 million business establishments operating in Canada. Businesses with fewer than 100 employees represented 97.6% of all firms and they employed more than 6.8 million Canadians, equal to about 64% of all private sector employment. Public Works and Government Services Canada (PWGSC) estimates that SMEs account for 45% of the country's GDP, making their market value approximately $688.7 billion.[36] Small businesses represent 85% of Canadian exporters and generate export revenues of over $123 billion, or 35.9% of the country's total export value. There is strong potential for future export growth from this sector.[37]

SMEs *and Society*

It is indisputable that the sheer volume of firms and their economic impacts make SMEs vital to the communities in which they operate. Therefore it is axiomatic that these business organizations have enormous potential to influence social sustainability and citizen well-being in Canada. But they are not without their unique challenges and barriers to success.[38]

There is often a close relationship between small firms and their local communities, in part because they tend to be owner-operated, they employ local people, and their functions directly impact the surrounding area.[39] The owners' and employees' values may therefore be important motivating factors for social engagement. SMEs tend to engage in informal social responsibility activities that are more difficult to measure[40], but it is possible they will be more likely than larger companies to voluntarily engage in social activities due to their close community links. Many traditional incentives for social responsibility do not appeal to SMEs because they are not affordable or attainable. Short-term financial drivers often hold more weight for small firms. Certification standards are increasingly being relied upon for supply chain management; however, the cost and bureaucracy required to attain the required standards can be so onerous that SMEs choose not to pursue them.[41] In order to take advantage of the positive opportunities that SMEs offer Canadian (and global) society, policies and resources need to be developed that clearly articulate the business case for sustainability, and that are articulated in a language that business owners are able to understand.

Here we may draw a parallel between SME take-up of sustainable business practices and the potential benefits of adopting information and communications technologies (ICTs). Both are perceived to have potential productivity benefits for firms; the former often being related to eco-efficiencies and customer loyalty, and the latter to transformation through organizational change and business processes. Research indicates that SMEs tend to invest in ICTs less than large firms,[42] just as they are less likely to participate in environmental improvement programs or to invest in capital-intensive upgrades.[43] In both cases, many SMEs have not been persuaded by the business case for greater investment, presumably because they are not convinced of the predicted financial returns. Ironically, many recommendations for SME engagement in sustainable development refer to developing linkages between firms[44] and online resources that provide a "one-stop-shop" of information.[45] But without effective use and knowledge of ICTs, these recommendations become much less attainable and/or applicable. To add to the problem, lower productivity leads to decreased competitiveness, which perpetuates the cycle of short-term focus on survival versus a longer-term focus on sustainable development.

SMEs and the Environment

Environmental organizations such as the Canadian Centre for Pollution Prevention (C2P2), the Canadian Environmental Technology Advancement Centres (CETACs), and some University centres, including the Dalhousie University Eco-efficiency Centre, have focused their efforts on the greening of SMEs. But despite an acknowledgement that these firms are important contributors to Canada's economy and environment, there remains a lack of available data to quantify their success. Earlier in the paper we referenced some troubling trends on greenhouse gas (GHG) emissions for Canadian industry as a whole. But even today it is not clear exactly how much SMEs contribute to the GHG problem and other pollution challenges, or whether large firms really do account for the vast majority of pollution as the media often portrays. A number of studies have been conducted with mixed results that attempt to answer that question, two of which we describe below.

In 2002 the Ontario region of Environment Canada (EC) contracted the Ontario Centre for Environmental Technology Advancement (OCETA) to produce a report estimating the toxic outputs of the province's SMEs. Findings were generated at a facility-level and research tested the hypotheses that up to 80% of Ontario's facilities have fewer than 500 employees, and that they are responsible for 80% of pollutants and waste produced. OCETA used EC's National Pollutant Release Inventory (NPRI) to test the hypotheses and found that 87% of the 912 facilities they reviewed had fewer than 500 employees. Furthermore, they revealed that 62% of pollutants came from these facilities. While this number was lower than expected, it was still deemed a significant portion of Ontario's total environmental impact.[46]

A second study was done by Markus Biehl of York University and Robert Klassen of the University of Western Ontario. Their research was also based on the EC NPRI and targeted the manufacturing sector. The report was produced for Industry Canada in response to the need for a detailed inventory of SME pollutants in order to understand the role of SMEs in sustainable development. The level of analysis used by Biehl and Klassen was the firm, rather than the facility level. They argued that because a large firm can have numerous facilities, the method used by OCETA did not necessarily reflect the true SME community. The issue is that large firms often have greater resources to invest in capital and environmental management systems; therefore, comparing independent facilities with ones that are parts of larger entities is unfair. This change in definition resulted in the conclusion that a somewhat lower proportion of pollutants is derived from SMEs than OCETA claimed. When taking into account only reporting facilities in the NPRI, SMEs were responsible for 16% of total pollution – a proportion much lower than the 62% found by OCETA in 2002. Biehl and Klassen also made estimates to account for non-reporting firms, but this only increased the result to 17%. In contrast, they found that small and medium-sized facilities owned by large firms contributed 32% of total pollution.[47] This discrepancy in findings further points to the need for accurate data collection pertaining to SMEs.

SME Contribution to Pollution Abatement in Canada and Worldwide

As part of the effort to articulate the relationship between Canadian SMEs and the environment we felt it was also important to note the positive contributions of SMEs operating in the "environment industry" to sustainable development. At a time when emphasis is being placed on the potential role for environmental technologies and services in cleaning the environment it may be especially salient to look at the current contribution of SMEs to this activity.

The Canadian services sector has represented more than 45% of the economy's GDP for the past 20 years, and between 2000 and 2005 it grew significantly more than any other industry sector.[48] Elsewhere in the paper we have identified a lack of Canadian service sector representation in the international sustainable development arena. This gap represents an opportunity that may be filled to a significant degree by the SME community, given the importance of both the services sector and SMEs to the Canadian economy.

Since 1995 Statistics Canada has conducted an Environment Industry Survey in order to measure the size of the industry in terms of employment and economic output. According to the survey: "Environmental goods and services are used to measure, prevent, limit or correct environmental damage (both natural or by human activity) to water, air, soil as well as problems related to waste, noise, and ecosystems."[49] It is apparent from the survey results dating between 1995 and 2004 that the environment industry has been generating increasing revenues at a steady rate. In fact, during that time revenues increased from roughly $10.2 billion to $18.4 billion. In 2004 firms with less than 500 employees accounted for 89.5% ($16.4 billion) of the total revenues. Businesses with fewer than 100 employees were responsible for 62% ($11.4 billion). Thus, we may observe that the impact of SMEs on revenue generation in the environment industry is much higher than it is in the economy at large. This indicates that efforts targeted at these firms may have a greater return on investment than those aimed at the SME population in general, as well as increased capacity to prevent and remedy environmental challenges.

Exports in the environment industry were tracked from 1996 to 2004 and they also increased over time from $768 million to $1.5 billion. Again, SMEs accounted for the overwhelming majority of these revenues. Firms with fewer than 500 employees accounted for 85% of export revenues in 2004 and those with fewer than 100 employees earned 54% of the total. Based on the assumption that these technologies and services represent a key component for solving and/or mitigating environmental problems, SMEs could be an increasingly influential driver for Canada's success in reducing its ecological footprint domestically and assisting other countries in doing the same internationally.

Ecological modernisation (EM) theory[50] supports the idea that environmental protection can act as a source of innovation. The theory predicts new market opportunities for the environment industry to drive economic growth and emphasizes that the market is the best medium for solving environmental problems though promoting a voluntary approach to environmental innovation.[51] It is possible to apply this theory to the emergence and growth of environmental technologies since innovation has led directly to the development of new techniques to solve environmental problems. However, research in the UK by Revell and Blackburn has revealed that EM theory has flaws when applied to SMEs because these firms often do not fully support and/or understand the usual "business cases" for sustainability. Further, SMEs fear a loss of competitiveness, they lack access to resources and support systems, and voluntary approaches often encourage them to ignore sustainability issues. These findings are important considering the high percentage of SMEs that comprise the environment industry.

SME Motivation for Ecological Responsiveness

Earlier in the paper we described the known motivating factors for adopting sustainability practices and strategies in larger firms. Great care must be taken in extrapolating from these observations to the motivations of SMEs. However, evidence does exist which provides some insight.

The most wide-reaching survey of SME perspectives on the environment is produced by the Canadian Federation of Independent Business (CFIB). Its October 2007 publication revealed that of the survey's 10,800 respondents, 78.9% believed that it is possible to grow the economy while protecting the environment. 83% indicated a high level of individual concern, yet over a third expressed a need for more information about what could be done in their own business. The most salient issues for SMEs were the recycling of materials, energy conservation, and clean water/sewage.

By applying the CFIB data to Bansal and Roth's model of ecological responsiveness,[52] we may hypothesize that Canadian SMEs have the potential for high levels of responsiveness, but that barriers prevent them from fitting into a specific profile. This is consistent with the findings of Côté et al,[53] which revealed significant unfulfilled opportunities for improvement in the environmental management of SMEs. Despite a high level of individual concern, issue salience is often low due to a lack of available resources or basic understanding of the issues.

Despite high individual concern amongst Canadian SMEs, many still fail to take advantage of the strategic niche position as a green alternative. This may be due in part to the fact that regulations can stifle innovation and productivity. The cost and paperwork required for SMEs to adhere to regulations and certification standards may discourage them from making improvements to their operations. In Canada, the lack of clarity about which level of government is responsible for each regulation also causes uncertainty for SMEs. This is exemplified by the fact that 10% of all SMEs and 16% of firms in business for less than one year do not know if they are even subject to environmental regulations.[54] Resource limitations such as access to time, skills, and capital can also prevent the type of long-term strategic thinking that is necessary to obtain advantages from environmental innovation. Thus, in terms of environmental management SMEs are often reactive and tend to be driven by factors like cost and legislation, rather than by long-term strategic advantage or innovation. Further research including sectoral case studies would be required to confirm these hypotheses.

It is clear that in order for SMEs to increase their ecological responsiveness, incentives need to be clarified and barriers need to be addressed. More information about the environmental and social impacts of SMEs, coupled with improved information resources for firms, may help to increase perceived issue salience. Improved partnership opportunities and support systems between firms, as well as information offered by industry-related organizations, could enhance the legitimacy of actions adopted by SMEs.

DISCUSSION AND RECOMMENDATIONS

In the introduction to this paper we noted the removal of seven well known corporations from their former Canadian owners, and thus potentially from a leadership role in the discourse around business sustainability in Canada. Not necessarily coming from a sustainability perspective, financial services chiefs like Gordon Nixon (Royal Bank), Dominic D'Alessandro (Manulife) and others have expressed public misgivings about public policies that allow such free and easy takeover of Canadian firms by foreign owners.[55] D'Alessandro summed up his concerns in an address to shareholders thus: "I sometimes worry that we may all wake up one day and find that as a nation, we have lost control of our affairs."[56] Peter Munk, Chairman of the Toronto-based, world number one gold mining firm Barrick is also well-known for his patriotic assessment of the problems in his own industry: "[Swiss-based] Xstrata didn't exist six years ago. Now look what they've got. It requires balls, it requires guts, it requires vision. And those are not qualities that come to [our] senior corporate managers."[57]

Others, including the Investors Group (which is part of IGM and indirectly owned by Power Corporation) have refuted the "hollowing out" hypothesis. They cite Statistics Canada, TSX and CIBC World Markets data to argue that foreign ownership of formerly Canadian firms does not necessarily harm the climate for inward investment, that the value of the Toronto stock exchange (prior to the 2008 crash) was at its highest value ever with 20% more firms listing in the last five years, and that the reasons for the high level of foreign merger and acquisition activity in Canada in 2006 and 2007 are not necessarily an absence of vision or important bodily parts. Instead they were: i) high commodity prices; ii) record corporate profits (in addition to strong balance sheets, low debt and rising cash flow in Canadian firms); iii) the maturation of Canadian firms into global players that attract more attention from foreign interests; and iv) the cash rich status of potential buyers and/or their access to cheap cash for purchasing Canadian firms. Similarly, in a 2007 report from the Ontario government funded Institute for Competitiveness and Prosperity's *Agenda for Canada's Prosperity* the case was made that between 1985 and 2006 the number of Canadian firms that could be described "global leaders" on the basis of sales and international market share criteria increased from 33 to 72 despite the disappearance of established icons like Falconbridge and Inco and budding superstars like ZENON.

Thus, from a corporate sustainability perspective, the issue may not be whether the supposed hollowing out phenomenon is good or bad for the Canadian economy. Mining and energy companies (and in due course our banks when they finally lose their regulatory shelter) may forego their Canadian-owned status, but does that matter if they retain a commitment to Canadian and international sustainability? Perhaps it may even be argued that if Brazilian, Russian, Indian and Chinese firms buy our sustainability-minded mining and energy firms and US banks buy our socially responsible financial institutions it may even help green those sectors in other jurisdictions on the "trojan horse" principle?[58] It is even possible that Inco may learn more from Companhia Vale do Rio about sustainability than the other way around. The problem with this reasoning is of course the question of leadership and institutional forces. Earlier in this paper we laid out what institutional drivers are germane to corporate greening and we argued in constructing our index and exploring SME motivations that a clear and specific business case, external signaling, and delivering performance are all important to the dynamic for change. Based on the index developed for this paper, if those drivers are primarily jurisdictional rather than sectoral, transfer of ownership to the Netherlands, Sweden or the UK might be a net good thing, but transfer of ownership to the Russian Federation or the US would seem less reason for celebration. If a firm is transferred in ownership to Moscow or New York it will be subject to fewer institutional pressures for signaling sustainability and that may in turn lead to a diminution of commitment in the Canadian subsidiaries rather than the reverse.

From a Canadian perspective, it may be futile (and perhaps even damaging to the economy in the long term) to try and prevent international acquisitions on grounds of chauvinism or narrow patriotism, even if those sentiments were rooted in sincere and honourable commitments to sustainability by Canadian policy-makers and opinion-leaders. A rather more interesting approach might be to follow the rising tide of emerging global leaders in Canada and explore the extent to which their domestic and international interests in sustainability might be re-enforced through jurisdictional and sectoral institutional forces. Such an approach has been explored for energy, food and ICT sectors in Canada[59] where, coupled with compelling and appropriate business cases, intangible institutional assets such as leadership capabilities, reputation, brand and social capital were all deemed vital to achieving commercial advantage and international competitiveness within a sustainability framework.

According to the Institute for Competitiveness and Prosperity, the sectors that are producing the greatest numbers of new global leaders in Canada today include high-tech (e.g. information and communication technology and life sciences), services (e.g. retail and financial services), and manufacturing. So if we follow this logic, the challenge now is how to ensure that firms like Research in Motion, Nortel, Open Text, Thomson Corporation, Jean Coutu Group, Empire Company Ltd, George Weston Ltd, McCain's, Magna, Bombardier, Nova Chemicals, Onex, Manulife (and many other financial institutions) can play a Canadian leadership role as internationally connected, competitive, and outward-facing, self-confident commercial enterprises. In the boxes below we identify some signs of progress in both RIM and George Weston Ltd, at least at the level of the adoption of a broader vision. If these and other companies build sustainability criteria more actively into their business models for reasons of international competitiveness, that may also help inspire their Canadian-based peers and supply chain partners – including SMEs – to deliver an equivalent level of competitiveness and international commercial focus whilst driving a sustainability agenda, both in signaling and performance terms.

And yet if we look for evidence of these firms and even their sectors on the basis of signaling, we see relatively little evidence of their international sustainability engagement today. The presence of large Canadian companies in sustainability membership organizations like WBCSD and the Global Compact is still heavily dominated by the energy and resource industries. If we remove Alcan (now taken over by RioTinto), and Hudson's Bay (now US owned), the active membership of these bodies (as of late 2007) featured just a handful of large Canadian firms: Barrick, BCE, BC Hydro, Bombardier, Catalyst Paper, Enbridge, Nexen, Petro-Canada, Suncor Energy, Talisman, Teck Cominco and TransAlta, all but three of which are in energy supply or resource management. And virtually every Canadian GRI compliant report in the GRI database is from a financial institution or from a firm in the extractive or energy industries. This all translates into a significant threat – and a significant opportunity – for Canadian business positioning on sustainable enterprise.

RESEARCH IN MOTION (RIM) LTD based in Waterloo, Ontario is an example of a successful, international Canadian high-tech firm and the Blackberry, its signature product, is helping people move towards a paperless workplace. James (Jim) Balsillie is RIM's co-CEO and in 2002 he founded the CENTRE FOR INTERNATIONAL GOVERNANCE INNOVATION (CIGI), a non-profit, non-partisan research institution based in Waterloo. CIGI addresses international governance challenges including those relevant to sustainable development. For example, its 2007 work program included an event titled "*Towards Sustainable Energy Futures?*" that brought together leading experts and policymakers to discuss issues related to energy and the role of international governance institutions in environmental sustainability.

www.cigionline.org/

To summarize:

1 Corporate Canada can claim a respectable mid-table position on international sustainability engagement and performance compared with other G8 nations; however:

2 At the present time, corporate Canada is engaged in the international sustainability agenda primarily through sectors that may migrate into foreign ownership (resource-based industries) and or that have a largely domestic focus (energy supply and financial institutions).

3 Today, corporate Canada is largely unengaged in international sustainability fora and initiatives through sectors such as high technology, manufacturing and services that may expect to assume increasing economic prominence both domestically and internationally in future years.

4 SMEs in Canada are often overlooked in the discussion of Canadian business and sustainability, only referenced when someone important to their industry tells them they should register for ISO 14001 or embrace social responsibility.

5 Nevertheless, SMEs have a significant economic, social and environmental impact in Canada in their own right, and if properly incentivized may be expected to contribute significantly to domestic and (through export sales) to international sustainability, especially in environmental services and technologies.

LOBLAW COMPANIES LTD is owned by George Weston Ltd and is Canada's largest retailer and one of the country's biggest private sector employers. It is an example of a non-resource-based company that is starting to lead the corporate sustainability movement in the service sector in Canada. Loblaw's mission includes its responsibilities to *respect the environment* and *make a positive difference in our community*. By sourcing from Canadian providers of agricultural produce, prioritizing organically certified supplies, and reducing the carbon footprints of its facilities, Loblaw is emerging as a sustainability leader in Canada. Loblaw is working with industry partners to develop an industry-wide carbon footprint measure and increasingly offers product lines that promote healthier eating and environmentally-responsible lifestyles.

www.loblaw.ca/

From this we may infer:

1 There is a strong incipient risk of Canada losing ground in our league table of corporate engagement and performance, both through loss of Canadian corporate membership and through non-engagement of key sectors;
2 There is a need for initiatives to promote institutional leadership on sustainability and competitiveness in currently non-active key sectors if Canada wishes to avoid progressive marginalization on corporate and SME sustainability issues.
3 There is a need to find ways to involve SMEs through promoting their involvement in market-based opportunities in newer and increasingly important sectors of the Canadian economy e.g. high technology, services (including retail and finance) and manufacturing (including environmental technologies).

CONCLUSION

The kind of intervention that would make a real and immediate difference to the engagement and performance of corporate Canada in sustainable development would be if the CEOs of the large Canadian corporations we listed earlier (i.e. Research in Motion, Nortel, Open Text, Thomson Corporation, Jean Coutu Group, Empire Company Ltd, McCain's, Magna, Bombardier, Nova Chemicals, Onex, Manulife, and other financial institutions) followed the lead of Galen Weston Jr., quoted at the beginning of this paper. Further speeches and commitments made by the CEOs of oil and gas, mining and power companies will continue to be welcomed by commentators in Canada. But these industries are already in the game and the institutional and legitimization pressures surrounding them are probably sufficient to ensure further progress, regardless of domestic and foreign ownership structures. The really urgent challenge for Canada therefore – and also the major opportunity – is for corporate leaders in the high technology, services and manufacturing industries of the future to initiate change and embrace the best of Canadian SME innovators as their partners in driving sustainability agendas for commercial advantage both at home and abroad. Pursuing and legitimating these opportunities, anchoring them in Canadian values, and securing public policy support should now be a high priority for anyone who cares about Canada and what the country's leading businesses and their SME partners can contribute to domestic and global sustainability.

This will require many more Canadian businesses and their leaders to recognize their roles both as value-creators and social actors. It will require them to take a stance and carry their sectors both domestically and internationally, for that is how our SMEs can also maximize their contributions and win in international markets too. And it will require governments at the federal and provincial levels to create the right conditions for success, through their own procurement practices and regulatory activities, rewarding firms large and small with market opportunities and sustainability pay-offs, rather than focusing exclusively on compliance or bureaucratic registration processes.

The 2008 credit crisis and the proposed response of the incoming US Administration to regenerate the US economy by spectacular investments in the green economy provide a powerful signal that recent events in the financial sector may further the arguments made in this paper. A similar vision, and similar investments are now required in Canada.

NOTES

1 Throughout this paper we accept the Brundtland definition as described in the World Commission on Environment and Development (WCED) Report, *Our Common Future* (Oxford University Press, 1987).

2 Natural Resources Canada (NRCan), *The State of Energy Efficiency in Canada* (2006).

3 NRCan's definition of the industrial sector includes all manufacturing and mining activities, forestry and construction; but it excludes electricity generation and non-industrial sectors and services.

4 Environment Canada, *Canada's 2005 Greenhouse Gas Inventory: A Summary of Trends* (2005). Available at: www.ec.gc.ca/pdb/ghg/inventory_report/2005/ 2005summary_e.cfm; accessed October 2, 2007.

5 E.C. Kurucz, B. Colbert, and D. Wheeler, "The Business Case for Corporate Social Responsibility," in D. Seigel, A. Crane, D. Matten, J. Moon, and A. McWilliams, eds. *The Oxford Handbook of Corporate Social Responsibility* (Oxford University Press, 2007).

6 J.A. Aragón-Correa and S. Sharma, "A contingent resource-based view of proactive corporate environmental strategy," *Academy of Management Review* 28 (2003), 71–88.

7 E.A. Rubio-Lopes and J.A. Aragon-Correa, "Proactive corporate environmental strategies: myths and misunderstandings," *Long Range Planning* 40 (2007), 357–81.

8 S.L. Hart, "A natural-resource-based view of the firm," *Academy of Management Review* 20 (1995), 986–1014; S.L. Hart, "Beyond greening: Strategies for a sustainable world," *Harvard Business Review* 75 (1997), 66–76; T.N. Gladwin, J.J. Kennelly, and T.S. Krause, "Shifting paradigms for sustainable development: Implications for management theory and research," *Academy of Management Review* 20 (1995), 874–907; P. Shrivastava, "The role of corporations in achieving ecological sustainability," *Academy of Management Review* 20 (1995), 936–60.

9 D. Esty and A. Winston, *Green to Gold: How Smart Companies Use Environmental Strategy to Innovate, Create Value, and Build Competitive Advantage* (Yale University Press, 2006); P. Hawken, A. Lovins, and L.H. Lovins, *Natural Capitalism: Creating the Next Industrial Revolution* (Little, Brown and Company, 1999); C.O. Holliday, S. Schmidheiny, and P. Watts, *Walking the Talk: The Business Case for Sustainable Development* (Greenleaf Publishing, 2002); B. Willard, *The Sustainability Advantage: Seven Business Case Benefits of a Triple Bottom Line (Conscientious Commerce)* (New Society Publishers, 2002); B. Willard, *The Next Sustainability Wave: Building Boardroom Buy-in* (New Society Publishers, 2005).

10 S. Sharma and H. Vredenburg, "Proactive corporate environmental strategy and the development of competitively valuable organizational capabilities," *Strategic Management Journal* 19 (1998), 729–53; S. Sharma, A.L. Pablo, H. Vredenburg, "Corporate Environmental Responsiveness Strategies," *Journal of Applied Behavioral Science* 35 (1999), 87.

11 T. Berkhout and I. Rowlands, "The voluntary adoption of green electricity by Ontario-based companies: The importance of organizational values and organizational context," *Organization & Environment* 20 (2007), 281.

12 I. Henriques and S. Sharma, "Pathways of stakeholder influence in the Canadian forestry industry," *Business Strategy and the Environment* 14 (2005), 384–98; S. Sharma and I. Henriques, "Stakeholder influences on sustainability practices in the Canadian forest products industry," *Strategic Management Journal* 26 (2005), 159–80.

13 P. Bansal, "Evolving sustainably: A longitudinal study of corporate sustainable development," *Strategic Management Journal* 26 (2005), 197–218.

14 A.J. Hoffman, "Institutional evolution and change: Environmentalism and the US chemical industry," *Academy of Management Journal* 42 (1999), 351–71.

15 P.A. Adler and S.W. Kwon, "Social capital: Prospects for a new concept," *Academy of Management Review* 27 (2002), 17–40; R.E. Freeman, *Strategic Management: A Stakeholder Approach* (Pitman, 1984); R.E. Freeman and J. Mcvea, "A Stakeholder Approach to Strategic Management," in: M.A. Hitt, R.E. Freeman, and J.S. Harrison, eds. *Handbook of Strategic Management* (Blackwell, 2001), 189–207; J. Nahapiet and S. Ghoshal, "Social capital, intellectual capital and the organizational advantage," *Academy of Management Review* 23 (1998), 242–66.

16 R.V. Aguilera et al. "Putting the S back in corporate social responsibility: A multilevel theory of social change in organizations." *Academy of Management Review* 32 (2007), 836–63.

17 D. Etzion, "Research on organizations and the natural environment, 1992–present: A review," *Journal of Management* 33 (2007), 637–64.

18 M. Biehl and R. Klassen, "How Much Pollution Do Canadian Small- and Medium-sized Enterprises REALLY Generate? An Empirical Analysis of Pollution through the Manufacturing Sector" (Industry Canada, 2006); R. Coté, A. Booth, B. Louis, "Eco-efficiency and SMEs in Nova Scotia, Canada." *Journal of Cleaner Production* 14 (2005), 542–50.

19 C. Egri, et al. "Managerial perspectives on corporate environmental and social responsibilities in 22 countries," *Academy of Management Proceedings* (2004), C1–C6; E. Reynaud, et al. "The differences in values between managers of the European founding countries, the new members and the applicant countries: Societal orientation or financial orientation?" *European Management Journal* 25 (2007), 132–45; B. Scholtens and L. Dam, "Cultural values and international differences in business ethics," *Journal of Business Ethics* 75 (2007), 273.

20 T. Maak, "Responsible leadership, stakeholder engagement, and the emergence of social capital," *Journal of Business Ethics* 74 (2007), 329–43; D. Wheeler and J. Thomson, "The New Bottom Line: Energy and Corporate Ingenuity," in A. Heintzman and E. Solomon, eds. *Fuelling the Future: How the Battle over Energy Is Changing Everything* (Anansi Press, 2003), 242–93; B. Colbert, E. Kurucz, and D. Wheeler, "Building the Sustainable Organisation Through Adaptive, Creative Coherence in the HR System," in R. Burke, and C. Cooper, eds. *Building More Effective Organizations: HR management and Performance in Practice* (Cambridge University Press, 2008).

21 Both the Netherlands and Sweden have four corporations listed in the 2007 Corporate Knights "Global 100' Most Sustainable Corporations in the World" ranking that is compiled using data from Innovest Strategic Value Advisors. This compares with the much larger economies of Canada, France, and Germany which have five, six, and five corporations respectively. See www.global100.org/index.asp; accessed 7 October 2007. In the "Business Action" component of the 2007 AccountAbility Responsible Competitiveness Index, Sweden scored first, the Netherlands scored sixth, and Canada 13th. See www.accountability21.net/publications.aspx?id=878; accessed 13 October 2007). From a broader jurisdictional sustainability perspective, it is interesting to note that as a country Sweden scores particularly well on sustainability metrics. For example in the Yale Centre for Environmental Law and Policy Environmental Sustainability (ESI) and

Environmental Performance (EPI) indices. See www.yale.edu/epi/ and www.yale.edu/esi/; accessed 13 October 2007). Sweden ranked 2nd and 4th respectively in the 2006 pilot EPI and 2005 ESI against Canada's 8th and 6th places. Sweden also scored 1st place in the World Conservation Union (IUCN) 2001 combined Human and Ecosystem Well-being Index, compared with Canada's 7th place. See www.iucn.org/en/news/archive/2001_2005/press/wonrank.doc; accessed 13 October 2007.

22 See www.globescan.com/sose_overview.htm; accessed 7 October 2007.

23 D. Wheeler and A. Tamlyn, "Canadian Business and the Sustainability Challenge: Engagement and Performance," Working Paper (Dalhousie University, 2008).

24 G. Kell and D. Levin, "The global compact network: An historic experiment in learning and action," *Business & Society Review* 108 (2003), 151–81.

25 D. Wheeler and M. Ng, "Organisational Innovation as a Driver of Sustainable Enterprise. Standardization as a Potential Constraint," in S. Sharma and M. Starik, eds. *Stakeholders, Environment and Society: New Perspectives in Research on Corporate Sustainability* (Edward Elgar, 2004), 185–211.

26 S. Waddock, "Creating corporate accountability: Foundational principles to make corporate citizenship real," *Journal of Business Ethics* 50 (2004); A. Willis, "The role of the global reporting initiative's sustainability reporting guidelines in the social screening of investments," *Journal of Business Ethics* 43 (2003).

27 S.M. Livesey and K. Kearins, "Transparent and caring corporations? A Study of sustainability reports by the body shop and Royal Dutch/Shell," *Organization and Environment* 15 (2002), 233; P. Bansal and G. Kistruck, "Seeing is (not) believing: Managing the impressions of the firm's commitment to the natural environment," *Journal of Business Ethics* 67 (2006), 165–80.

28 M. Lopez, A. Garcia, and L. Rodriguez, "Sustainable development and corporate performance: A study based on the Dow Jones Sustainability Index," *Journal of Business Ethics* 75 (2007), 285–300.

29 P. Cerin and P. Dobers, "What does the performance of the Dow Jones Sustainability Group Index tell us?" *Eco-Management and Auditing* 8 (2001), 123–33.

30 Wheeler and Tamlyn, "Canadian Business and the Sustainability Challenge" (2008).

31 Geometric means are more favourable for a ranking of this nature because of possible skewing of arithmetic means by outlying data. In this case the use of arithmetic means resulted in only one substantive change in the overall ranking, moving Italy to 7th place rather than joint 6th with Canada.

32 There are discrepancies in how organizations define business size; Statistics Canada uses a revenue-based approach, while others such as the Canadian Federation of Independent Business (CFIB) refer to the number of employees in a firm. These discrepancies can hinder data accuracy and ease of comparability; however, they generally yield the same results. For the case of this paper we will define SMEs as having fewer than five hundred employees. Small businesses will be categorized by having fewer than one hundred employees.

33 M. Schaper, "The Challenge of Environmental Responsibility and Sustainable Development: Implications for SME and Entrepreneurship Academics," in U. Fuglistaller,

H.J. Pleitner, T. Volery, and W. Weber, eds. *Radical Changes in the World – Will SMEs Soar or Crash?* (University of St. Gallen, 2002).

34 P. Bansal and K. Roth, "Why companies go green: A model of ecological legitimacy," *Academy of Management Journal* 43 (2000), 717–36.

35 T. Fox, "Small and Medium-Sized Enterprises (SMEs) and Corporate Social Responsibility: A Discussion Paper" (2005). Available at: www.iied.org/SM/CR/documents/CSRandSMEs.pdf; accessed 22 September 2008.

36 Public Works and Government Services Canada (PWGSC), "Acquisitions on the Internet: Importance of SMEs" (2006). Available at: www.pwgsc.gc.ca/acquisitions/text/sme/importance-e.html; accessed 1 October 2007.

37 Industry Canada, "Small Business Research and Policy: Key Small Business Statistics" (July 2007). Available at: www.ic.gc.ca/epic/site/sbrp-rppe.nsf/en/rd02189e.html; accessed 2 October 2007.

38 It is important to note that less than half of new companies established between 1993 and 2003 survived beyond their third year.

39 Fox, "SME's and Corporate Social Responsibility" (2005).

40 CBSR, "Engaging Small Business in Corporate Social Responsibility: A Canadian Small Business Perspective on CSR" (2003). Available at: www.cbsr.ca/files/ReportsandPapers/EngagingSME_FINAL.pdf; accessed 19 September 2008.

41 CFIB, "Achieving Eco-prosperity: SMEs' perspectives on the environment" (Canadian Federation of Independent Business, 2007).

42 R. Martin and J. Milway, "Enhancing the Productivity of Small and Medium Enterprises through Greater Adoption of Information and Communication Technology" (Information and Communications Technology Council, 2007).

43 Schaper, "The challenge of environmental responsibility and sustainable development" (2002).

44 Fox, "Canadian Business and the Sustainability Challenge" (2008).

45 CFIB, "Achieving Eco-prosperity" (2007).

46 OCETA, "Analysis of Toxic Pollutant Loadings from Ontario Small- to Medium-Sized Manufacturers Reporting to the National Pollutant Release Inventory" (OCETA, 2002).

47 Biehl and Klassen, "How much Pollution" (2006).

48 Industry Canada, "Service Industries: Services Sector Overview – October 2006." Available at: www.ic.gc.ca/epic/site/si-is.nsf/en/ai02201e.html; accessed September 22 2008.

49 Statistics Canada, "Environment Industry: Business Sector 2002 (revised) and 2004," 5, catalogue 16FOO08XIE (Industry Canada, 2007).

50 EM theory was most famously promoted by Porter and van der Linde (1995) in a landmark article published in the *Harvard Business Review* and is predicated on the concept that the environment and the economy can benefit simultaneously from such innovation, most notably at the large firm level.

51 A. Revell and R. Blackburn, "SMEs and Their Response to Environmental Issues in the UK," *Occasional Paper Series* 57 (Kingston Business School, 2004).

52 Bansal and Roth, "Why companies go green" (2000).

53 Coté, Booth, and Louis, "Eco-efficiency and SMEs" (2005).

54 CFIB, "Achieving Eco-prosperity" (2007).

55 "Is corporate Canada being 'hollowed out'?" *CBC News* (2007). Available at: www.cbc.ca/news/background/mergers/hollowed-out.html#excerpt; accessed 2 October 2007.

56 Address to shareholders given 3 May 2007 by Dominic D'Alessandro, president and chief executive officer of Manulife Financial Corp (cited in CBC, 2007).

57 D. Olive, "Canada beating industrial retreat – Foreigners have snapped up nearly 600 Canadian firms since the start of 2006" (2007). Available at: www.thestar.com/article/ 213327; accessed 6 October 2007.

58 The late Anita Roddick, founder of The Body Shop International, claimed this as her reasoning for the sale of her stake in the company to L'Oréal (interview with Claudia Cahalane for *The Guardian Unlimited*, 3 November 2006. Available at: http:// business.guardian.co.uk/story/0,,1938850,00.html; accessed 6 October 2007.

59 Wheeler and Thomson, "The New Bottom Line" (2003); D. Wheeler and J. Thomson, "Brand Barons and the Business of Food," in A. Heintzman and E. Solomon, eds. *Feeding the Future: From fat to famine – how to solve the world's food crisis* (Anansi Press, 2004), 194–235; D. Wheeler, K. Thomson, M.A. Perkin, "Sustainability, Social Capital and the Canadian ICT Sector," in G. Toner, ed. *Sustainable Production: Building Canadian Capacity* (University of British Columbia Press, 2006).

9 A Child of Brundtland: The Institutional Evolution of the National Round Table on the Environment and Economy

SERENA BOUTROS

The National Round Table on the Environment and Economy (NRTEE) is an institution tasked with no small feat "to identify, explain and promote the concept of sustainable development for all sectors and regions of the nation."[1] This chapter will begin by tracing the genesis of the NRTEE and go on to examine the institutional evolution the NRTEE has undergone since its inception. The chapter will discuss both the failures and successes of the NRTEE and conclude with a brief look at what lies ahead.

BACKGROUND: EVENTS LEADING UP TO THE CREATION OF THE NRTEE

The government of Canada worked actively to bring the World Commission on Environment and Development (WCED) to Canada. During the WCED's visit to Canada in 1986, its most memorable meeting was with the Canadian Council of Resource and Environment Ministers (CCREM). In direct follow-up to the special meeting of the WCED with the CCREM in Edmonton, the National Task Force on Environment and Economy (NTFEE) was formed. The purpose of the NTFEE was to initiate dialogue between Canadian environment ministers, senior executive officers from Canadian industry, Environmental Non-Government Organizations (ENGOs) and academe regarding the integration of the environment and the economy. The type of multi-stakeholder dialogue the NTFEE attempted to initiate was one of the first attempts of its kind; however, the stage had been pre-set in Canada during the 1980s by the achievements of other multi-stakeholder consultation process used to develop the Workplace Hazardous Material Information System

(WHMIS) and the Canadian Environmental Protection Act (CEPA).[2] Concurrently, industry became increasingly concerned that Canada would follow the lead of the US and adopt a strong regulatory approach to address environmental issues and goals. There was also escalating public concern regarding the environment and public support for increased environmental enforcement and regulation. It was against this backdrop and largely at the initiative of the Manitoba government that the NTFEE was created.

In addition to the government of Canada's efforts to bring the Brundtland Commission to Canada, it eagerly sought to host one of the four global conferences leading up to the release of the Commission's report, *Our Common Future*. The Canadian conference was viewed as the most successful of the four; in the view of the Brundtland Commission, the Canadian conference was a major breakthrough, so much so that Canada was asked to host the World Conservation Strategy Conference a few months later. In the wake of these successful conferences, the federal government sought to sustain the momentum generated by the Brundtland Commission within Canada and Canada's position as an international leader on the sustainable development front, and thus encouraged and supported the NTFEE (through the federal Department of the Environment). Furthermore, the CCREM was in the midst of debating its future role and was therefore open to innovative ideas such as creating the NTFEE.[3] Indeed, the history and accomplishments of the NTFEE laid the groundwork for the establishment of the NRTEE.

One of the most important outcomes of the NTFEE was the "social learning" of its participants which generated an enormous amount of mutual respect between industry and environmental participants, a respect that had previously been absent. This respect was built upon the environmentalist's realization that industry representatives were well-informed, concerned with environmental issues and willing to consider alternative points of view. In turn, industry representatives discovered that environmental representatives respected industrial perspectives and wanted to seek out solutions which avoided unnecessary damage to industrial interests. Furthermore, it was the first time that government ministers directly associated with senior industry decision-makers, such as CEOs, in a neutral setting. Previous contact of this sort had either never occurred or was riddled with conflict over regulatory and legislative initiatives or environmental problems that had to be resolved. The NTFEE provided a neutral forum where ministers and industry representatives could begin to understand each other, build mutual respect and communicate effectively. It also sent a signal to other cabinet ministers that environment ministers were becoming important economic players, thereby enhancing the profile of the environment portfolio.

The NTFEE recognized that the demand for information and involvement in social, economic and environment problems was rapidly growing and as such, government and decision-making processes would have to change to meet this demand. Further, the NTFEE recognized that single points of view made in isolation cannot sufficiently reflect the complexities inherent when interests pertaining to the environment and economy are involved and therefore recommended a new process of consultation. This new process of consultation it envisioned would bring together individuals who exercised considerable influence over policy and planning decisions as well as those who could bring information and different perspectives to the debate, thereby mirroring closely the structure of the NTFEE itself.

Additionally, the NTFEE envisaged that these consultations would be premised on consensus building and intended to exert direct influence on policy and decision-makers at the highest levels of government, industry and non-governmental organizations. In pursuit of these objectives, the NTFEE recommended the establishment of roundtables on the environment and economy. The NTFEE encouraged each province and territory to create its own roundtable comprised of senior decision makers, who have extensive interest and expertise in environmental and economic issues, and that it report its conclusions not only directly to the first ministers of their jurisdictions, but also to the public. Furthermore, the NTFEE proposed the establishment of a national roundtable which would draw its membership from the provincial and territorial roundtables, providing a national perspective on regional environment and economic issues. The national roundtable was to also have additional members from the federal cabinet and national NGOs, as well as national labour and business associations. The recommendation to create roundtables was the most salient recommendation emanating from the report of the NTFEE. The task force experience revealed the merits of the multi-stakeholder consultation process and sought to institutionalize its approach.

THE NATIONAL ROUND TABLE ON THE ENVIRONMENT AND THE ECONOMY

A roundtable is defined as a multipartite body, designed to reflect diverse backgrounds and experiences, differing perspectives and insights, as well as divergent values and beliefs.[4] As such, members are drawn from senior levels of government, the industrial/corporate sector, academe, NGOs such as research institutes, professional associations, public interest groups and the scientific community. Above and beyond attempting to mirror the success the NTFEE enjoyed through its roundtable experience, the planning committee of the NRTEE sought out the roundtable structure so that:

They [the NRTEE] could operate in the context of common imperatives-those being the challenge of integrating environment and economy in our institutions and forms of decision-making, and the need to share across sectors the responsibility for bringing about that change ... through their members and their respective spheres of influence, they act as catalysts to forge new strategic partnerships, to stimulate the search for viable solutions, and to build a broad consensus on what must change, who should bear the costs, and how and when those costs should be born.[5]

The thrust behind the creation of the NRTEE was to engage Canadian society with the ideas and principles of SD. The roundtable structure is most fitting as the characteristics of SD issues are such that they are beyond any one sector, jurisdiction or community. They are inherently inter and trans-disciplinary and highly normative and therefore, demand innovative ways to identify solutions, partnerships and policy directions. The roundtable structure was adopted because "It ... tries to reach across all institutional lines, be they governmental, business, occupational, social, political, environmental, or regional, in order to encourage the flexibility of responses necessary for the transition to a sustainable society."[6]

MAJOR EPOCHS/LEADERS/PLAYERS

The NRTEE was announced as "an independent, multi-sectoral body that will seek to foster environmentally sustainable economic development in Canada ... providing leadership in the new way we must think about the relationship between the environment and the economy and the new way we must act ... will forge new ideas and new partnerships necessary to address important issues."[7] The NRTEE has undergone various iterations since its inception and its institutional history has been characterized by constant changes in the interpretation of its mandate and the processes it uses to identify and analyze issues. Indeed, a key question it has grappled with since its inception is what role(s) it should emphasis in striving to fulfill its very broad mandate?[8] It is not that the mandate of the NRTEE has changed over time, but rather the emphasis the NRTEE has placed on different aspects of its mandate has shifted.

There are three possible roles suited to the NRTEE, in terms of fulfilling its broad mandate, an advisory role, a catalytic role and an advocacy role. An advisory role would entail the NRTEE offering advice to cabinet ministers, senior officials and the prime minister. In order for the NRTEE to effectively sustain an advisory role the advice must be seen as highly credible and non-partisan. A catalyst role involves inciting others to undertake desired actions or changes in behavior, stimulating others to undertake new initiatives, sharing information and experiences and forging new partnerships. The catalytic role entails isolating and building awareness around an issue so that others see the urgency of action. Lastly, an advocacy role was put forward; however, it was believed by many from the outset that pursuing a strong advocacy role is in some respects contradictory to the very nature of the NRTEE. Typically advocacy bodies are founded by like-minded individuals with similar interests, while the NRTEE sought to foster dialogue and consensus seeking among those with conflicting perspectives on issues that cut across boundaries and jurisdictions. Furthermore, in order for the NRTEE to become a responsible advocate, it would have to invest heavily in becoming a centre of expertise and information in specific areas,[9] an objective the NRTEE did not have in mind initially. In order to effectively examine the various stages of institutional development the NRTEE has undergone, its history will be broken down into five time periods.

Period I: Institutional Birth (1988–1990) [10]

The final report of the NFTEE recommended the creation of a national round-table and Prime Minister Brain Mulroney announced the creation of the NRTEE, through an Order in Council, in 1988. In October of that year, a planning committee was struck by the NRTEE's executive team. The executive team was lead by Dr. David Johnston, the first chair of the NRTEE and two federal civil servants, Ann Dale and Dorothy Richardson. The planning committee included the former secretary-general of the WCED, Jim MacNeill, former Quebec premiere, Pierre Marc Johnson, vice-president of Inco Inc., Roy Aitken, president of Dow Chemical, Dave Buzzelli and Judge Barry Stuart from the Yukon Territory. The committee was specifically chosen to mirror the spectrum of communities the NRTEE was seeking to draw upon.[11] The planning committee laboured over the terms, conditions and criteria for appointment to the NRTEE as well as identified possible candidates to recommend to the Prime Minister's Office for appointments. The planning committee concluded that the NRTEE should have 24 members (serving two year appointments), meet four times a year and should draw its membership from the following four general groupings: political; business/industry; science/strategic policy; and public interest/labour/professional.

The NRTEE was to operate under the context of common imperatives, spreading across sectors the responsibility to integrate the environment and the economy into Canadian institutions and decision-making forums. The NRTEE was not intended to develop or deliver programs, nor was it attempting to become a major source of expertise on technical aspects of the economy or ecological systems. Rather, the NRTEE sought, through its members, to act as a catalyst, forging new strategic partnerships and building broad consensus on the implementation of sustainable development in Canada:

The National Round Table has a special catalytic role to play in Canadian society in identifying, explaining and promoting the concept of sustainable development for all sectors and regions of the nation. The National Round Table will:

Address the social issues of environment and economic development by bringing together the best information and analysis possible;

Advise the Prime Minister and the federal government on ways and means to integrate environmental and economic considerations in decision-making processes, beginning with the recommendation of the National Task Force on the Environment and Economy; and on Canada's role concerning global issues of environment and sustainable development;

Promote and catalyze awareness and understanding throughout Canadian society of the cultural, social, economic and policy changes required to shift to more sustainable forms of economic development;

Forge new partnerships in Canada that are necessary to resolve issues and overcome barriers for the achievement of sustainable development.[12]

This mandate drove membership appointments from the highest levels of decision-making in the country. It also led the planning committee to deviate from the recommendation of the NTFEE that it follow the traditional hierarchal federal/provincial alignment. There was no explicit attempt to seek provincial representation; rather, members who represented critical networks and constituents were sought. Nevertheless, there was an emphasis placed on ensuring regional, gender and ethnic representation; in fact, nine of the twenty-three members were women, a significant feat at the time.

The emphasis on a catalytic role was further evidenced by the membership criteria espoused by the NRTEE's planning committee. Members were to be chosen based on their ability to influence their respective constituents and broader networks, "the ability of the roundtable to exert wide influence is based upon ... the credibility of the people making up the roundtable and the ability of these individuals to gain access to the key sectors of our society."[13] It is also important to note that the idea of the NRTEE pursuing an advisory role was not well-received during this period. While the reluctance of the NRTEE to focus on an advisory role stemmed largely from the fact that federal cabinet ministers were members,[14] the planning committee made it quite explicit that it did not want the NRTEE to become "just another" federal advisory body.[15] Pursuing an advocacy role was hardly considered; given the broad range of adversarial perspectives being brought together for the first time it was easy to see that the NRTEE could not have agreed on enough to serve as an advocate for any issue.

Establishing the internal processes of the NRTEE consumed most of this period as there were no precedents for the NRTEE to follow since roundtables were a novel concept. The plenary organized the NRTEE's internal processes into the form of standing committees. The majority of the action took place within the committee construct; essentially the committees were microcosms of the NRTEE with the NRTEE acting like the board of directors for each committee. The five committees operated with a fair degree of autonomy, in terms of developing projects, undertaking analysis and presenting their reports to the plenary for approval. It is important to note that during the

NRTEE's first year of operation, its membership meetings were completely closed. The plenary felt having closed meetings was critical to ensuring frank and unfettered peer-to-peer dialogue, especially in light of the fact that ministers of the crown were members of the round table.[16]

The planning committee was arguably the most salient internal factor during this period as it affected all aspects of the NRTEE (i.e. its mandate, internal processes and operations). The planning committee, lead by the former principal of McGill University and first chair of the NRTEE, Dr. David Johnston, was made up of several experienced and senior officials from all spectrums of Canadian society. They were intent on sustaining the momentum generated by the WCED and were seeking innovative approaches to the implementation of SD in Canada. As the NRTEE was not only one of the first of its kind in Canada but also around the world, and Dr. Johnston was seen as highly credible by his political superiors, very little political direction was given to the planning committee in terms of establishing the model the NRTEE would follow, what it should do, how it would operate, and its expected results. It was this high degree of flexibility coupled with the planning committee's thirst for innovation which allowed the NRTEE to flourish into the first expanded model of an arms length body with government representation, including the federal ministers of the Environment, Finance, Industry, Science and Technology and Energy, Mines and Resources. Significantly, the inclusion of ministerial representation at the table established creditability for the NRTEE among senior business leaders and high-profile environmental leaders.

While free trade issues monopolized the government's agenda during this phase, the NRTEE did enjoy a very high public profile. It was able to maintain a high public profile due to the following external factors: its members were appointed by the prime minister and it was one of the only bodies in Canada that reported directly to the prime minister; there was much enthusiasm for the roundtable movement as an appropriate institutional response to the WCED; there was much interest from other countries in the roundtable process and since Canada was considered one of the few countries in the world to embrace roundtables, the NRTEE was gaining credibility in the international arena; and, over 200 roundtables had been established in Canada during this period.

During this period work was underway to develop a federal environmental strategy, and more broadly a federal sustainable development strategy. The NRTEE side-stepped this issue as it was not convinced of the merits of a federal sustainable development strategy given the vast regional, geo-physical, social and economic differences in Canada.[17] Nevertheless, when a coalition of ENGOs and Aboriginal groups presented the prime minister with an environmental strategy, entitled *A Greenprint for Canada*,[18] the NRTEE quietly supported it.[19]

The NRTEE worked very closely with the provincial and territorial roundtables during this period and began to adapt the roundtable process to the municipal level. The NRTEE established the Canadian Association of Round Table Secretariats (CARTS) as an information sharing forum between the secretariats in order to avoid the duplication of projects and leverage any potential opportunities for collaboration. An opportunity to sustain Canada's position as a prominent international champion of SD issues was presented by the planning and preparations for the United Nations Conference on Environment and Development (UNCED) in 1992 in Rio de Janeiro. UNCED presented the NRTEE with the unique opportunity to assist the federal government in preparing for the pre-conference meetings as well as to develop four seminal policy papers to present at UNCED and to Prime Minister Mulroney.

Given the expectations placed on the NRTEE, its budget during this period was insufficient. At the time, budget pressures were so severe that the only way the NRTEE survived was through the experience and networks of those comprising the planning committee.[20] Moreover, the budget for the NRTEE came from Environment Canada, leaving the NRTEE with little to no direct control over its finances. Another challenge the NRTEE faced was securing high-level ENGO representation. There was a sentiment throughout some parts of the highly diverse ENGO community that the NRTEE would be "doomed to fail" and as such were reluctant to take part in the process. Some ENGOs such as Greenpeace, refused to join the NRTEE, however, "While not all environmentalists are avid participants, most are willing to sacrifice time and even reputation"[21] to be involved.

Period II: Fine Tuning (1991–1993)

This period was characterized by a focus on the internal processes of NRTEE, involving a fine tuning of its operations and method of organization. It reaffirmed its mandate to play a catalytic role and therefore began to raise its profile with the Canadian public. The NRTEE "conducted major studies in several key areas, sponsored many workshops and conferences and generally acted as a catalyst to promote the principles and practices of sustainable development."[22] Furthermore, the NRTEE sought to heighten its focus on the catalytic role by strengthening the plenary. While the NRTEE had split its focus into thirteen different areas, one of its foremost commitments during this period was analyzing various issues related to UNCED. The NRTEE tendered formal advice to the prime minister, the bulk of which was accepted, and sent two of its members, Pierre Marc Johnson and Jim MacNeill, as part of Canada's official delegation to Rio. As a direct follow-up to Rio, *Projet de Société – Planning for a Sustainable Future*, championed by Minister Jean Charest, was established in November 1992 and the NRTEE played a significant role in its coordination.

Alterations to the NRTEE's internal processes reflected the NRTEE's desire to enhance its role as a catalyst. In particular, the standing committee system was abandoned. Members began to question the effectiveness of the standing committee approach and there was a general feeling that they were operating too autonomously, so much so that it precluded members from engaging in real ongoing dialogue, missing the original intent of the NRTEE. Eliminating the standing committee system was to strengthen the role of the plenary, as the plenary was seen as the true point of "bridge building" across different perspectives. A task force approach was adopted in its place wherein the plenary was responsible for developing initiatives and assigning them to task forces, thereby shifting more decision-making to the plenary while allowing the task forces to act as a means of broadening engagement among members. This was also the first time that the NRTEE began developing real communication tools intended to broaden its audience. While thirty "working papers" covering a wide range of issues were commissioned, a coordinated communications function was not.

The internal factors affecting the changes during this period were predominately member-driven, the most significant of which was the resignation of cabinet ministers. Originally, meetings were closed to members only, as it was believed this was critical for frank and unfettered peer-to-peer dialogue. However, intense and persistent lobbying from various interest groups, especially by environmental groups, to open up the meetings and share the direct access to cabinet ministers prevailed and members were given the right to bring one resource person with them to meetings. Additionally, cabinet ministerial staff lobbied and received the right to bring resource people from both their political and bureaucratic staffs with the following result, "Meetings became a combination of open and closed sessions, with a trend to more and more open sessions, and a corresponding shrinking of frank and open dialogue, until after the second year, all meetings were open."[23] The consequences of this were that "ministers became increasingly uncomfortable with this extended dialogue, and with good reason … Confidentiality and discretion obviously became increasing concerns … Ministers did not sit at the table as individuals, they could not take off the mantle of the Crown … Tragically, the enlightened and frank dialogue that had been building through the first few years began to shrivel."[24] Soon after, cabinet ministers began to completely disregard the "no substitutes" policy[25] until eventually cabinet ministers resigned. As a consequence of the departure of cabinet ministers, Ann Dale and Jim MacNeill, the original planners of the NRTEE also resigned. A new Chair was appointed, George Connell, former president of the University of Toronto and Ronald Doering took the position as Executive Director. Doering pushed for a more decentralized approach to the internal operations of the NRTEE and was a formidable force behind the NRTEE receiving a legislated mandate.

This period was characterized by a marked interest in the environment, especially by the federal government post-Rio. The NRTEE took advantage of the opportunity to make significant pre and post-Earth Summit contributions, as Rio represented a landmark meeting resulting in breakthrough treaties such as the UN Convention on Biological Diversity and the UN Framework Convention on Climate Change. At the same time, there was an impression that the NRTEE was becoming unfocused and lacked effective organizational coherence. Some felt that the NRTEE was overextending itself, tackling rather ambitious social change initiatives and issues that were too large and open-ended. Consequently, the NRTEE's focus on the "economy" side of issues was slipping and so was the interest of the business community. Moreover, some members were beginning to feel discontent with the NRTEE's preoccupation with achieving consensus on every single issue. Members felt that this overriding need to achieve consensus detracted from honest dialogue and disagreement and forced the NRTEE to use "lowest common denominator" language in its reports.[26] The resignation of cabinet ministers reflected a near breakdown of the original design and intent of the NRTEE. The original planners of the NRTEE intended to break traditional institutional molds by creating an expanded decision-making model with the most senior levels of government as participants.[27] When this failed, those involved with the original planning of the NRTEE lost hope in their vision that the NRTEE would break traditional institutional molds and foresaw its lapse into becoming "just another advisory body."[28]

Period III: Growing and Developing (1993–2000)

The most significant change was securing a legislative base. *The National Round Table on the Environment and Economy Act* received royal assent by the House of Commons in 1993 and for the first time in Canadian history a roundtable became a departmental corporation with a legislated mandate. The Act enshrined a far-reaching mandate for the NRTEE, including both a catalytic role and a more focused advisory role, section 4 of the Act reads:

4. The purpose of the Round Table is to play the role of catalyst in identifying, explaining and promoting, in all sectors of Canadian society and in all regions of Canada, principles and practices of sustainable development by

(a) undertaking research and gathering information and analyses on critical issues of sustainable development;

(b) advising governments on ways of integrating environmental and economic considerations into their decision-making processes and on global issues of sustainable development;

(c) advising those sectors and regions on ways of incorporating principles and practices of sustainable development into their activities;

(d) promoting the understanding and increasing public awareness of the cultural, social, economic and policy changes required to attain sustainable development; and

(e) facilitating and assisting cooperative efforts in Canada to overcome barriers to the attainment of sustainable development.[29]

How does the legislated mandate differ from the NRTEE's previous mandate, as affirmed at the time of its institutional birth? The NRTEE's original mandate contained a reference that it advise the prime minister – this was omitted and replaced with "advising governments" and "sectors and regions." Also, the reference to "critical issues of environment and economic development" were replaced with "sustainable development." This begs the question, was the term sustainable development injected to strengthen the purpose of the NRTEE or was the removal of the reference to "economy" a deliberate attempt to cloud the mandate of the NRTEE? Likewise, was the omission of the NRTEE "providing advice to the prime minister" an attempt by the bureaucracy to limit the NRTEE's access to the center of power, or was it simply to broaden the focus from just the federal government to "all governments and sectors?"[30]

Throughout this period the NRTEE steadily shifted away from a catalyst role, and began to move away from reaching out to wider networks and audiences with its reports, to primarily reaching out to the federal government in the form of providing formal advice.[31] NRTEE members were becoming increasingly reluctant to return to their spheres of influence to share the NRTEE's advice and recommendations, "Some members said they didn't feel confident enough about the subject to undertake any kind of outreach or communications."[32]

The NRTEE sought to narrow its focus during this period. While a wide range of issues were still put forward, the new Chair (Stuart Smith) and new President & CEO (David McGuinty) strove to reduce the number of issues the NRTEE members worked on at any one time (they thought that thirteen was far too many). Working toward this end, the NRTEE began to organize itself around "State of the Debate" reports.[33] The State of the Debate reports: "Synthesize the results of stakeholder consultations on potential opportunities for sustainable development. They summarize the extent of consensus and reasons for disagreements, review the consequences of action or inaction, and recommend steps specific stakeholders can take to promote sustainability."[34]

The introduction of 'State of the Debate' reports was a significant change as it did not require members to attain consensus on every issue, but rather sought to focus on where consensus could not be reached, clearly explaining where and why there was disagreement. Relaxing the requirement that members achieve consensus on every issue and recommendation placated the increasing number of members who became frustrated with constantly striving to reach agreement.

The key internal factors responsible for the NRTEE receiving a legislated mandate were the workings of the chair and president & CEO at the time,

George Connell and Ronald Doering. When Connell approached Doering offering him the position as executive director of the NRTEE, Doering was reluctant to accept the position in fear that the NRTEE would not continue to survive unless it received a legislated mandate. Therefore, he made this a condition of his accepting the position. This drove Connell to forcefully sell the idea that the NRTEE needed legislative status to the minister of Environment, Jean Charest. He was able to do so, and even though there was a change in government that same year, he was able to successfully re-convince the new Chrétien government, allowing the *National Round Table on the Environment and Economy Act* to receive royal assent in 1993.

Paradoxically, optimism in the roundtable movement was plummeting during the mid-90s, as by 1995 most of the 200 roundtables throughout Canada had been dissolved or were winding down. Furthermore, the NRTEE was struggling to stay on the radar of the PMO and the senior bureaucracy against the backdrop of program review and increasing budget constraints. Given the demise of most of the roundtables throughout Canada during this time, as well as the collapse of various Canadian advisory councils,[35] the legislation was critical to the survival of the NRTEE.[36] As the period wore on, the NRTEE built credibility as an independent source of advice to the prime minister and the federal government and in 1996 the NRTEE's Greening the Budget submissions began. These annual submissions served to further enhance its creditability and increased its profile within the federal bureaucracy.

Period IV: High Productivity (2001–2005)

This period saw the NRTEE undertake leading-edge work on high profile issues such as climate change, brownfields, urban sustainability, and ecological fiscal reform. The NRTEE's ability to play an independent advisory role for the federal government was signaled by references made to the NRTEE in the federal budget and the increased importance placed on the annual "green budget" consultations it conducted with the Department of Finance. Interestingly, there was also a modest effort to move beyond an advisory role and keep the catalyst function alive by attempting to increase its follow-up efforts on its State of the Debate reports. This is best illustrated through the NRTEE's work on brownfields, notably, its efforts to extend its work "beyond" Ottawa and serve as a catalyst for change: "Previously, our efforts did not extend much beyond the development and delivery of the analysis, information, insights and recommendations contained in our State of the Debate reports ... we now recognize that we need to raise the level of understanding of these issues more broadly ... such awareness raising is essential to generate momentum for these changes and to effectively influence decision-making."[37]

The NRTEE narrowed its focus and decreased the number of issues it examined, as stated by David McGuinity, "Today, we take a more strategic approach in our work, using an environmental scanning process to select a much smaller number of issues."[38] In limiting the issues it examined, there was a renewed determination to only consider issues that were central to the environmental-economic challenges facing Canada where the NRTEE could make a unique and valuable contribution. Additionally, the NRTEE began to organize its work in terms of *strategic outcomes for the organization*, rathr than *activities for individual programs*. These strategic outcomes were aimed at producing new, relevant, credible knowledge, working partnerships, broad awareness and consideration of the NRTEE's findings by government, industry and key decision-makers.

These changes were driven by Stuart Smith in his role as the Chair. Smith had very strong political connections[39] and used these to work "behind the scene" to increase the profile of the NRTEE within the federal government. Smith encouraged senior officials and Ministers to make reference to the NRTEE where the NRTEE believed it could make a useful contribution (i.e., climate change, brownfields). Smith and McGuinty moved the NRTEE further in the direction as an independent advisor to the federal government by reducing the number of issues examined and raising the profile of the NRTEE within the federal government: "References for the NRTEE from the government rarely happen on their own. They usually reflect considerable background preparatory work by the NRTEE senior staff and Chair to brief senior officials and Ministers on an issue and propose the NRTEE as an effective independent, credible source of advice."[40]

Two crucial external factors affected the NRTEE during this period. The Treasury Board Secretariat of Canada (TBS) heightened its corporate accountability and reporting requirements (as stipulated in the Reports on Plans and Priorities, Departmental Performance Reports, and Program Activity Architectures). This presented a host of challenges for the NRTEE as its capacity to increase its level of reporting was not matched with the finances to do so. Second, environmental issues triggered increased political interest as climate change concerns gained traction throughout Canadian society and internationally. The increased reporting requirements coupled with unanticipated requests from the federal government concerning advice on climate change exacerbated budgetary pressures. The NRTEE had to prioritize the issues it would examine in order to find the funding to support these newfound obligations. Importantly, as the NRTEE received these requests from the government for advice, members were becoming increasingly concerned about the independence of the NRTEE. Some members even began to feel that the NRTEE should not be responding to requests from the government.[41] The growing number of requests for advice from the bureaucracy caused internal struggles within the membership regarding how to strike a proper balance between taking requests for advice from the government while still maintaining the NRTEE's independence from government.

Period V: Re-Assessment (2005–Present)

The ongoing dynamic between the NRTEE's pursuit of a catalyst role and an advisory role is perhaps most prevalent during this current period. While the NRTEE still predominately fulfills an advisory role, an excerpt from its 2005 Workplan and Performance Reports reveals efforts to move beyond providing advice to the government: "Beyond the delivery of advice to the Prime Minister and the federal government, the overriding goal will be to reach opinion leaders and influencers in all jurisdictions, as well as community, business and civil society leaders, to build a constituency of support for the NRTEE recommendations and further work that evolves from our next phase."[42] The NRTEE sought to strike a balance between producing advice and promoting advice: "In the past few years, the NRTEE has tended to focus its efforts on producing advice. And our tack record has been highly successful in this regard. However, we recognize that this represents only half the equation. As we go forward into the next year, we plan to shift the balance in favour of promoting advice, particularly to stakeholder groups who historically have seemed less aware of our work."[43]

A major impact was a change in government in February 2006 to the Stephen Harper-lead Conservative party after 13 years of Liberal rule. The environment was not on the agenda of the newly elected government, particularly climate change issues. Secondly, under the newly elected government the Privy Council Office changed the reporting structure of the NRTEE to the minister of the Environment, from the prime minister. Lastly, there was an even greater emphasis placed on transparency and reporting requirements across all government departments and agencies. The new requirement, to report *against* government-wide priorities and strategic outcomes, was introduced and continued to put pressure on the capacity of the NRTEE to fulfill these obligations. This also meant that the NRTEE had to introduce new priorities for itself (i.e., government-wide priorities rather than just its internal priorities such as advancing sustainable development in Canada), "The NRTEE's second priority is to strengthen management accountability and systems and to implement new federal initiatives ... to ensure that the stewardship of its financial and human resources is effective and aligned with government-wide initiatives."[44]

While the environment was not on the agenda of the Harper government when it took office in the beginning of 2006, within a year it had risen to the top as public concern skyrocketed. The NRTEE was also becoming an authoritative voice on climate change issues. Indeed, the Harper Cabinet turned to the NRTEE for advice concerning its Kyoto obligations, medium and long-term greenhouse gas (GHG) emission scenarios and climate change adaptation strategies.[45] Specifically, the Conservatives requested the NRTEE's advice concerning targets and scenarios for reducing Canada's emissions of GHGs in the medium and long-term term. In response, the NRTEE released, *Advice on a Long-term Strategy on Energy and Climate Change for Canada* in 2006 and its *Interim Report to the Minister of the Environment on Medium- and Long-term Scenarios for GHG Emission and Air Pollutant Reductions* was released in 2007.[46] Furthermore, the *Kyoto Protocol Implementation Act*, which received royal assent in 2007, requires the federal government to produce annually, a Climate Change Plan and Statement- detailing the measures and policies it will put forward to fulfill its emissions reductions as per the Kyoto Protocol. Subsection 10(1) of the *Kyoto Protocol Implementation Act*[47] stipulates that the NRTEE must provide a response to the government's plan by assessing the likelihood that the government's Plan and Statement would achieve its stated objectives.[48] Hence, because of political drivers the thrust of the NRTEE's focus in this current period is on climate change issues.

This period had its share of stresses including relying on Acting Presidents from 2002– 2007. The increased reporting requirements imposed by the TBS via the *Federal Accountability Act* put further pressure on the NRTEE's already constrained capacity and budget. Additionally, the change in the reporting structure of the NRTEE, from the prime minister to the minister of the Environment was seen by some members as an institutional downgrade of the NRTEE,[49] causing some members to question the future course of the NRTEE.

SUMMARY OF THE NRTEE'S INSTITUTIONAL EVOLUTION

Breaking down the various iterations the NRTEE has undergone through the five distinct time periods has illuminated some very interesting trends regarding the ongoing dynamic in its mandate/focus, the evolution of its internal operations, the relevance of internal and external factors as well as the barriers, challenges and opportunities the NRTEE has faced over the years.

Mandate: From Catalyst to Advisor

At the time the NRTEE was conceived, it was strongly believed that it was well-positioned to use its credentials and networks of members to take on the challenging role of catalyst.[50] It was also believed that there was a strong need for such a catalyst because SD issues transcend boundaries and jurisdictions. The NRTEE was perceived to be the ideal forum to bring together players who are traditionally at odds with one another to work together to communicate and identify key issues concerning the environment and the economy and work within their respective spheres of influence to forge new strategic partnerships. As one of the only bodies in Canada reporting directly to the prime minister, the NRTEE was in an ideal position to fulfill this catalytic role. The hardnosed efforts on the part of the planning committee to carve out a new and innovative niche for the NRTEE drove the NRTEE to interpret its mandate through a broad catalyst lens during its first years of operation. As the focus of the NRTEE was predominately on fulfilling a catalyst role during its formative years, any attention to the NRTEE providing an advisory function was downplayed. Moreover, cabinet ministers sat as members of the NRTEE and as such did not feel comfortable being a part of a body which made public recommendations to the government.

The NRTEE's emphasis on an advisory role emerged in the mid-1990s and is still dominant. This is not to say that regard for a catalyst function has been abandoned; the two roles are not mutually exclusive – providing advice is a large part of fulfilling a catalyst role. For example, the NRTEE did play a catalyst role in respect to particular issues, such as urban sustainability, brownfields, and sustainable development indicators, while at the same time voicing its intent to broaden its role as an advisor to the federal government.[51] What explains the shift in the NRTEE's focus and interpretation of its mandate from catalyst to advisor? Evolutions in the NRTEE's internal processes and operations as well as crucial internal and external factors played a fundamental role.

Evolutions in the NRTEE's internal processes and operations were, in large part, driven by the need to sustain the active engagement of its members and to ensure that they were directly involved in developing and approving issues and program recommendations. Throughout the various iterations in its internal operations, from standing committees to task forces to plenary-only approaches, no one method has proved to satisfy all members. In every period, members expressed frustration at not being able to be more involved and "having time to only 'rubber stamp' recommendations."[52] This need to actively engage members spurred a majority of the changes to the internal processes of the NRTEE.

INTERNAL FACTORS AND INFLUENCES

There are a few key factors which stand out as significantly affecting the way in which the NRTEE interpreted its mandate. The Chair of the NRTEE is one of the most important internal factors shaping the mandate/focus of the NRTEE, particularly in respect to the NRTEE's role as advisor. The personal styles, preferences, interests, and connections, have shaped the NRTEE's mandate/focus over the years. In the NRTEE's formative years, specifically during its birth, the credibility of David Johnston, allowed for the NRTEE to develop with very little political oversight, which meant it had a great deal of flexibility when shaping its mandate and its internal processes. Toward the mid-1990s, the personal political and bureaucratic connections of Stuart Smith allowed the NRTEE to gain access to political decision-makers and raised the profile of the NRTEE within the federal government.

A second crucial internal factor was the working relationship between the chair and president and CEO. The dynamics between the chair and president and CEO have shaped the evolution of the NRTEE. For example, George Connell's support of Ronald Doering's decentralized approach allowed for several projects to be launched simultaneously. Doering's insistence that the NRTEE obtain legislative status led Connell to convince the minister of the Environment. Stuart Smith's political connections reinforced David McGuinty's desire to "work the Ottawa system"[53] increasing the NRTEE's access and influence at the federal level.

The members of the NRTEE undoubtedly played a crucial role in shaping the mandate of the NRTEE; the engagement of its members ultimately dictated the mandate of the NRTEE. Securing a high-level of commitment from its membership was key to ensuring that they would take the findings and recommendations of the NRTEE back to their respective spheres of influence and communities of interest in order to build the kind of awareness, partnerships, and networks that are central to a catalyst role. When the NRTEE experienced success as a catalyst, it was because a NRTEE member took a strong sense of ownership to the issue (for example the Chair of one of the Task Forces) and played an active role in taking the issue to broader audiences. Nevertheless, throughout the history of the NRTEE there has been reluctance from many members to take up this role and this has contributed to the NRTEE's propensity to take on an advisory role. Moreover, members had no control over appointments (members are appointed by the governor in Council), meaning there is no way to ensure that the membership is compatible or that newly appointed members will provide the broad cross-section of interest needed for the NRTEE to effectively tackle SD challenges.

The internal processes and operations of the NRTEE indirectly impacted the shaping of the NRTEE's mandate. For example, the planning process may have hindered the involvement of members thus causing them to feel disengaged and thus not willing to take on a catalytic role. Lastly, one of the most crucial factors affecting the NRTEE's interpretation of its mandate was and still continues to be its budget. The NRTEE has suffered chronic budgetary pressures since its inception and continues to face financial hardship especially in the current climate of increased accountability and reporting obligations. The capacity of the Secretariat is relatively small and the increased reporting burdens threaten its ability to fulfill its obligations. Arguably, its budget never matched its mandate to identify, explain and promote, in all sectors of Canadian society and in all regions of Canada, principles and practices of sustainable development.

External Factors

In the opinion of some, external factors have not played a significant role over the course of the organization's history. Since the NRTEE's inception, political interest in the environment has peaked and plummeted, and throughout both peaks and dips, the NRTEE had to work hard to maintain the interest of ministers and senior officials and to ensure its relevancy, "In short, the NRTEE has had to work hard for everything it has achieved in Ottawa over the years."[54] On the contrary, others have expressed the opinion that external factors are what matter most, especially public and political interest in the environment.[55] Nevertheless, it is clear that external factors are becoming increasingly important to the future trajectory of the NRTEE, especially in terms of its mandate. Currently, the most significant external factor affecting the NRTEE is the enhanced reporting requirements and the high standards of accountability all government agencies are being held to. The requirement that all federal departments and agencies must, on an annual basis, report how they have contributed to government-wide priorities will potentially constrain how the NRTEE interprets its mandate. While this increased accountability will not affect its position as a credible source of independent advice to the federal government, it could potentially preclude the NRTEE from expanding any catalyst efforts as any issue it would chose to undertake or any broader outreach activities would have to be directly linked to government-wide priorities.

SUCCESSES AND FAILURES

When contemplating the successes and failures of the NRTEE as an institution meant to engage Canada with SD, one must first take into account the fact that success and failure are subjective notions and measuring "soft processes" such as the outcomes of multi-stakeholder processes are very difficult and virtually impossible to quantify. In order to account for nineteen years of institutional changes, a set of five critical success factors relevant to a multi-stakeholder consensus-based body will be outlined and the NRTEE's performance against these success indicators will be measured.

1 Heterogeneous Membership

Heterogeneity within the membership of the NRTEE is an important factor for success as bringing together those with diverse experiences will allow a range of options and integrative solutions to be put forward and explored. A wide-ranging membership reinforces the democratic nature of reaching both points of consensus and overcoming conflict among different sectors of society. Furthermore, bringing together people from different sectors of society into one common forum allows for unusual networks, partnerships and alliances to form that otherwise may not have. The idea of having a high degree of heterogeneity among the members was to reach as many critical networks as possible. While the NRTEE has enjoyed a great degree of heterogeneity among its members, (more so in the early years of the NRTEE) it has been noted that the NRTEE has consistently lacked representation from the scientific community and civil society.[56] Moreover, appointments to the NRTEE are increasingly coming from the ranks of consultants. Consultants do not have a constituency per se and thus the fact that its current membership is increasingly comprised of consultants, throws the ability of the NRTEE to exert influence through the "spheres of influence" of its members into question.[57]

2 Seniority of Appointments

It is fairly obvious that a high level of seniority among the NRTEE members is a direct reflection of its success as an institution. Recently, the NRTEE has not faired well in regard to its seniority of appointments. This is especially true when considering that in its formative years, the NRTEE drew its membership from the highest levels of decision-making in the country. Furthermore, government is no longer represented at the table and members do not necessarily represent different sectors but are "educated laypersons in the field of sustainable development."[58]

3 Strategic Alliances/Partnerships

The NRTEE has constantly struggled to forge strategic alliances and partnerships. This battle is particularly true in recent times as the former interim president and CEO of the NRTEE noted, "forming strategic alliances is a big issue for the NRTEE right now … the NRTEE is in a strange place, it is quasi government, deals with the environment but also the economy … so it is very difficult to find the right partners."[59] A matter further complicating the ability of the NRTEE to forge strategic alliances is the degree to which this will affect its independence. The extent to which the NRTEE works with others directly affects its independence and as one of the foremost goals of the NRTEE is to be a source of credible and independent advice, this is of great concern. Nevertheless, the NRTEE also recognizes that the benefits of forging strategic partnerships must be weighted, such as increased outreach, cost-sharing, enhanced expertise. Hence, the issue of forming strategic partnerships is a matter of ongoing debate.

Perhaps one of the greatest failures of the NRTEE, true both in its formative years and still true today, is that it is fundamentally disconnected from and has never been able to strategically partner itself with the bureaucracy. "The NRTEE failed to realize it needed legs – the bureaucracy – it [the NRTEE] was too busy struggling to gain its independence."[60] There is also an overwhelming sense that the NRTEE never really had a chance at connecting with the bureaucracy, as in the opinion of some, "while the idea of the NRTEE was not opposed by the bureaucracy, so long as it did not have any real power and could be kept under their [bureaucratic] thumb."[61] Nevertheless, it is interesting to note that the bureaucracy had a greater propensity to "resent the advice of the NRTEE" and "saw the NRTEE as a threat"[62] during its early years when it had direct access to cabinet ministers, reported directly to the prime minister and when its emphasis on a catalytic role was most pronounced. More recently, the bureaucracy has been forced to reach out to the NRTEE for advice on substantial environment and economy issues, in particular with respect to climate change issues.

4 Self Organizing

Since its very inception, the NRTEE's roundtable process has been as unstructured as possible, giving members the flexibility to determine their own agenda, mode of operating and methods to reach their own conclusions.[63] The freedom to self organize is still apparent today as the current modus operandi of the NRTEE allows members to develop the major questions and issues that the NRTEE will focus on. The ability of the NRTEE to self-organize is considered a critical success factor as it fostered and continues to foster high levels of engagement by its members.

5 High levels of engagement by members

During the NRTEE's first years of operation, the level of energy and engagement by its members was phenomenal, "the dialogue between members during meetings was profound ... extraordinarily frank discussions and debates were had between members."[64] Nevertheless, sustaining this high-level engagement since its early years has been an ongoing struggle, "the biggest challenge facing the NRTEE by the end of the 1990s was sustaining the energy and engagement of senior members."[65] Sustaining the engagement of its members is crucial to the NRTEE's ability to exert influence over its member's spheres of influence in order to act as a catalyst for change.

6 Consensus building

The NRTEE was conceived as a consensus-based body, based upon the realization that if consensus on a recommendation could be reached it will have more relevance and legitimacy and a greater likelihood that it will be accepted by society in general. Hence, the NRTEE spent much of its early years highlighting the importance of consensus-based decision-making and putting mechanisms in place to ensure that achieving consensus was a paramount concern. While the NRTEE still attempts to indicate where consensus lies, analyzing the state of the debate in crucial areas can also highlight points of disagreement where differences exist. Continuing to seek out consensus where it exists is a good way for the NRTEE to proceed as consensus seeking is a preliminary exercise of democratic processes and will increase the legitimacy and likelihood that its recommendations will be accepted.

SO WHAT DOES IT ALL MEAN?

It has been the overwhelming response of many, from the original planners of the NRTEE, to former and current presidents and members, that the NRTEE has not been able to effectively engage Canada with SD nor has it reached its full potential. The NRTEE has experienced its fair share of challenges, such as: cabinet minister resignations; chronic budget/capacity pressures; a lack of strategic alliances/partnerships, especially with the bureaucracy; declining levels of seniority of appointments; changes to its reporting status, from the prime minister to the minister of the environment. In spite of all of this, the NRTEE has broken ground with its research and recommendations on certain issues, such as its work on brownfields, sustainable cities, sustainable development indicators, and most recently on climate change issues. Nevertheless, the NRTEE's greatest success is the fact that it is able to continue the roundtable movement, "setting the stage for a more robust position in the wider national debate."[66]

Although the NRTEE has transformed itself into primarily an advisory body, the roundtable movement it continues to symbolize has significantly affected how public involvement is done by governments throughout Canada. Through the NRTEE, it was successfully demonstrated that people from diverse backgrounds can be brought together to discuss and debate complex, and often thought to be irreconcilable, public policy issues, to come up with solutions and recommendations and profoundly influence the way public consultations have been embedded in governance in Canada. The roundtable movement that the NRTEE was central to and is one of the only remnants of, has changed the way public policy making is performed in Canada. Public/ multi-stakeholder consultations have been incorporated into most government procedures and are used at all levels of government to make recommendations on a variety of issues. Multi-stakeholder consultations are considered key public policy tools, shifting the policy and planning process to focus on engaging key stakeholders and citizens. The NRTEE effectively challenged the status quo on how decisions were made.

Moreover, the success of the roundtable process did not go unnoticed in the international arena. The NRTEE was highly sought after to share and teach other countries about its experience with this innovative institutional model.[67] The NRTEE successfully exported the roundtable process it developed and shared its experience with countries around the world, continuing the international struggle to ensure development that is sustainable. The overall influence of the roundtable process has been that "roundtables have shifted the historical relationship from hierarchy between governments and the business and environmental sectors to that of a network."[68] The NRTEE continues to demonstrate how complex issues related to sustainable development can be addressed by drawing together experts from all parts of Canadian society into an innovative institutional model.

CONCLUSION – THE ROAD AHEAD

They [roundtables] are small organizations with limited resources, bringing together – not people of like minds, but traditional adversaries. The scope of their task is nothing short of revolutionary. They require a paradigmatic shift in thinking and governing. They challenge nearly all the assumptions of classical economics. They threaten vested bureaucratic interests, in the public and private sector.[69]

Shifting through the various institutional changes to and interpretations of the NRTEE's mandate from its inception up to the present, it may be fair to say that perhaps the NRTEE has found its most natural fit – an independent advisor to the federal government. While it was never the original intention of the NRTEE's planning committee, an advisory role may be in the best interests of the NRTEE. The challenges in pursuing a catalytic role have arguably never been greater than they are in this current period. The increased emphasis on accountability and reporting obligations throws into question the ability of the NRTEE to take on a catalytic role. The requirement that all of the NRTEE's activities support government-wide priorities will only reinforce the importance of government references to the NRTEE and meanwhile the NRTEE must be seen as responding directly and immediately to government's questions and issues. Therefore, from a pragmatic and strategic perspective, the NRTEE would be wise to maintain its focus on fulfilling an advisory role. There is no doubt after examining the genesis and evolution of the NRTEE that it has indeed come a long way. The NRTEE has outlived the Science and Economic Councils of Canada and its hundreds of provincial, territorial and municipal roundtable counterparts to become one of the last remaining symbols of the roundtable movement in Canada.

NOTES

1 National Round Table on the Environment and Economy (NRTEE), *A Report to Canadians* (June 1991), 5–6

2 K. Ogilvie and B. Pannell, *Promoting Sustainable Development Using an Appropriate Mix of Policy Instruments, The Role of the NRTEE* (NRTEE Working Paper, 1995), 4.

3 Ibid., 5.

4 Ann Dale and Chris Ling, *The National Round Table on the Environment and the Economy: Expanded Decision-Making for Sustainable Development* (2005), 5.

5 Ibid., 6.

6 NRTEE, *Annual Report 1991–1992* (1992).

7 External Affairs Canada, *Address by the Right Honourable Brian Mulroney, Prime Minister of Canada, Before the UN General Assembly* (September 29, 1988).

8 Coopers & Lybrand Consulting Group, *National Roundtable on the Environment and the Economy: A Preliminary Discussion Paper* (NRTEE Working Paper, 1990), 6.

9 Ibid., 8.

10 This section draws upon the work of Tom Shillington, contracted by the NRTEE in 2006 to review the evolution of the mandate and planning and decision-making processes of the NRTEE.

11 Ann Dale, personal interview, 26 October 2007.

12 NRTEE, *A Report to Canadians*, 5–6.

13 NRTEE, *A Report to Canadians* (1990), 7.

14 The participation of cabinet ministers clouded the idea of the NRTEE providing advice to the government as ministers did not feel comfortable participating in a body that provided public advice to the government.

15 Dale, 2007.

16 The meetings were completely closed with the exception of Ann Dale and Dorothy Richardson. Dale, 2007.

17 Ann Dale, email correspondence, 1 May 2008.

18 Glen Toner, "The Green Plan: From Great Expectations to Eco-backtracking … to Revitalization?" in Susan Phillips, ed. *How Ottawa Spends 1994–95: Making Change* (Carleton University Press, 1995), 229–60.

19 In order to maintain the NRTEE's neutrality, its policy was to not endorse other people's work or documents. Dale, 2008.

20 Dale, 2007.

21 Ronald Doering, *Canadian Round Tables on the Environment and the Economy: Their History, Form and Function* (NRTEE Working Paper, 1993), 10.

22 Ibid., 6.

23 Dale and Ling, *Expanded Decision Making*, 6.

24 Ann Dale, "Multistakeholder Processes: Panacea or Window Dressing?" (unpublished, 1995).

25 Members were not allowed to send a substitute if they were unable to attend a meeting as the most important product of the roundtable process was the dialogue between members.

26 Tom Shillington, *Evolution of the NRTEE's Mandate and Planning Processes: Background Paper* (Shillington & Burns Consultants Inc., 2006), 12.

27 At some points, there were up to five cabinet ministers per meeting.

28 Dale, 2007.

29 Canada, *National Round Table on the Environment and the Economy Act* (1993), Section 4.

30 Shillington, *Evolution of the NRTEE's Mandate*, 13.

31 The exception was the "State of the Debate" reports, which could be seen as an attempt to reach those outside of the federal cabinet; however there was never a commitment of resources by the NRTEE to follow up on these reports.

32 Shillington, *Evolution of the* NRTEE*'s Mandate*, 14.

33 The NRTEE still operated through the use of Task Forces.

34 NRTEE, *State of the Debate Report: Water and Wastewater Services in Canada* (1996), 3.

35 The Science Council of Canada was created by federal statue and dissolved in 1993 and the Economic Council of Canada, a crown corporation, was also dissolved in 1993. Many still believe that gaining a legislated mandate was integral to the NRTEE surviving its provincial, territorial and municipal counterparts.

36 Ronald Doering, personal interview, 24 October 2007.

37 Message from the president and CEO, NRTEE *Report on Plans and Priorities* (2003–2004), 1.

38 Ibid.

39 Robert Slater, personal interview, 21 September 2007.

40 Shillington, *Evolution of the* NRTEE*'s Mandate*, 16.

41 Alexander Wood, personal interview, 18 July 2007.

42 NRTEE, *Workplan* (November 2005).

43 NRTEE, *Performance Report* (2005), 1.

44 NRTEE, *Report on Plans and Priorities* (2006–2007).

45 Glen Toner, "The Harper Minority Government and ISE: Second Year – Second Thoughts," in Glen Toner, ed. *Innovation, Science, Environment: Canadian Policies and Performance 2008–2009* (McGill-Queen's University Press, 2008), 3–29.

46 NRTEE, *Advice on a Long-term Strategy on Energy and Climate Change for Canada* (2006); and *Interim Report to the Minister of the Environment on Medium- and long-term scenarios for* GHG *emission and air pollutant reductions* (2007).

47 Canada, *Kyoto Protocol Implemenation Act* (2007).

48 NRTEE, *Response of the* NRTEE *to Its Obligations Under the Kyoto Protocol Implementation Act* (September 2007), 14.

49 Slater, 2007.

50 Interviews conducted by Coopers & Lybrand with NRTEE members, political staff and officials indicated that most emphasized the need for NRTEE to serve as a catalyst. Coopers & Lybrand Consulting Group (1990), 7.

51 NRTEE *Workplan* (2005); and NRTEE *Performance Report* (2005), 1.

52 Shillington, *Evolution of the* NRTEE*'s Mandate*, 19.

53 Ibid., 20.

54 Ibid., 21.

55 Ken Ogilvie, personal interview, 29 September 2007.

56 Dale, 2007; and Wood, 2007.

57 Dale, 2007.

58 Don Alexander et al., "The National Round Table on the Environment and Economy," *The Capable City* (The World Urban Forum, 2005), 31.

59 Wood, 2007.

60 Slater, 2007.

61 Doering, 2007.

62 Dale, 2007.

63 Alexander et al., *The Capable City*, 36.

64 Dale, 2007.

65 Dale and Ling, *Expanded Decision Making*, 8.

66 Ibid., 7.

67 Ronald Doering, (president and CEO of the NRTEE 1991–1995) constantly received invitations from countries around the world to share its round table experiences (Doering, 2007). Ann Dale (director of Internal Operations of the NRTEE 1989–1991) received requests from countries such as Norway and Thailand to share Canada's experience with the round table structure (Dale, 2007).

68 Alexander et al., *The Capable City*, 39.

69 Doering, *Canadian Roundtables on the Environment and the Economy*, 8.

10 The Best of Brundtland: The Story of the International Institute for Sustainable Development

LILLIAN HAYWARD

The Brundtland Commission's report *Our Common Future* provided a number of recommendations which, along with the highly popular definition of SD, led to visible institutional changes in Canada. Arguably the most successful of the institutional changes made in the wake of this report was the creation of the International Institute for Sustainable Development (IISD).

In the twenty years since the release of *Our Common Future*, this institution has had the opportunity to grow, mature and come into its own. Today, IISD is internationally recognized for its work. But, while IISD is a very successful institution, it is not entirely representative of how SD has fared in Canada. This juxtaposition of a successful organization within a struggling system provides a demonstration of the capacity of Canadian organizations when they are able to harness expertise and innovative thinking. IISD's experience is important because it has remained a strong and relevant organization in spite of potentially catastrophic organizational crises, risky endeavours and tumultuous times for the Government of Canada. This is a testament to IISD's strong and dedicated leadership which has been tested on many occasions only to emerge even stronger and more determined. Thus, while little has been done to significantly change the way we do things in the twenty years since the Brundtland Commission published *Our Common Future*, IISD is an illustration of the real potential of Canadian sustainable development institutions. IISD is internationally recognized for its work and this chapter will tell its story, evaluate its experience and look forward to the road ahead.

GENESIS: IDENTIFICATION OF DRIVERS

In the late 1980s the public interest in the environment was running very high, spurred on by several high profile environmental disasters both in Canada and abroad. It was in the midst of this atmosphere that the Brundtland Commission sent twenty-two men and women from its task force to Canada for a series of eleven high profile meetings held across the country.

Even before *Our Common Future* was published in 1987, the combination of public interest, heightened by the Commission's extensive Canadian tour and the impending release of the Commission's recommendations, caused governments in Canada to take action. One such action was the establishment of the National Task Force on Environment and Economy (NTFEE) by the Canadian Council of Resource and Environment Ministers. The NTFEE supported the main conclusions of the Brundtland Commission in principle and set out to address them in the Canadian context. Indeed, the final report of the NTFEE made a series of recommendations that reflected the relevance of the Brundtland Commission to Canada. It was from the report's significant international component that the idea for the creation of an international institute for sustainable development first emerged.

In 1988, Brian Mulroney was invited to address the United Nations General Assembly and was looking for a concrete commitment to include in his speech to his international counterparts. Given the immense public appetite for environmental action and the opportunity to address an international forum, it was a perfect time for Mulroney to announce the establishment of an institution with a focus on international environmental issues. The result was the announcement on 29 September 1988 by the prime minister to the United Nations General Assembly of the establishment of "a Centre which will promote internationally the concept of environmentally sustainable development. This centre will be located in Winnipeg and will work closely with the United Nations Environment Program and other like minded international institutions and organizations."[1]

Two years later, in March 1990, the Globe 90 Conference was held in Vancouver. At that time, it was one of the largest environmental conferences ever held, with over 2000 delegates attending and more than 600 exhibitors from 50 countries. Gro Harlem Brundtland provided a keynote address, stressing that "the [global environmental] crisis is a more real threat to the world than nuclear war, but unless the gap between rich and poor nations is bridged, it will to continue to grow."[2] The prevalence of international engagement in the over arching conference themes and the sheer number of international delegates, provided the perfect back drop for the signing of the funding agreement for IISD by Manitoba Premier Gary Filmon and Federal Environment Minister Lucien Bouchard. The agreement provided the new institution with $25 million over five years, funded by the Government of Manitoba, the Canadian International Development Agency, and Environment Canada.

Politics

The political landscape at the time of the creation of IISD was one in which environmental issues had gained significant momentum. In the first mandate of the Mulroney government, the environment did not figure prominently in a neo-conservative agenda focused primarily on economic issues such as free trade. The environment did surface at this time in the form of acid rain. It was dealt with more as a foreign policy issue and a focal point for Canada-US relations, but it did give the Prime Minister a familiarity with environmental issues. In addition, the Brundtland Commission's definition of sustainable development provided the kind of connection between the environment and the economy which was a better fit for the Conservatives than previous paradigms such as the Club of Rome's *Limits to Growth* study which called for conservation due to rapidly diminishing resources.

In spite of the new environmental paradigm being much better suited to the Conservative agenda, the federal government was generally anxious about the recommendations that would come out of the report. Public expectations were high and in order to pre-empt the advice that would come out of *Our Common Future*, the government initiated the NTFEE in 1986, following the Brundtland Commission's visit, in order to assess the relevance of the Commission's work for Canada.

The environment was an issue in the 1988 election campaign and Mulroney promised to deliver a strategy for the environment. Once re-elected, the 1989 Speech from the Throne emphasized the government's commitment to the environment through the recognition of environmental issues, strong support for the recommendations of the Brundtland Commission and the announcement of a new environmental agenda.[3] Major changes in the structure of government reflected this new agenda. Prior to this period the environment portfolio was a low profile assignment. As Mulroney noted: "It used to be considered a secondary or a tertiary assignment, with the Minister of the Environment reduced to mendicant status with a tin cup knocking on the door of the Minister of Finance to see if he would finance a program or two. We revolutionized that. We created the Cabinet Committee on the Environment to review the environmental implications of all government initiatives."[4] The environment portfolio experienced a rapid increase in importance thanks to the creation of the Cabinet Committee on the Environment. The minister of Environment was for the first time added to the powerful Cabinet Committee on Priorities and Planning.

This increased influence for the Department catapulted the new minister for the environment, Lucien Bouchard, a relatively new minister but one of Mulroney's close friends, into the inner circle of Cabinet. The result was positive from the perspective of environmental issues. However, this additional power lead to concerns that: "Departments would not be able to pursue their traditional mandates without Environment Canada's involvement. Many interpreted this shift in policy emphasis as interference in another department's affairs that would jeopardize their own autonomy."[5] Autonomy was a concern for departments, but so were budgetary constraints. These tensions mounted with the unveiling of Bouchard's major project and one which essentially defined this era of environmental policy-making within government, *Canada's Green Plan for a Healthy Environment*. The Green Plan was meant to be a master plan for the environment and sustainable development, but it was also a big budget item and there was concern that the development of this expensive plan in a time of fiscal constraint could result in additional burdens for all departments.

Before the Green Plan was even launched, Bouchard abruptly resigned from the government due to disputes that arose during the Meech Lake Accord process. Robert de Cotret, previously the President of the Treasury Board, was brought in to guide the Green Plan through a massive consultation process and substantial financial negotiations. The final budget for the Green Plan was $3 billion, which was shared amongst numerous government departments. Environmental issues receiving this kind of financial support increased the interest in the environment in Cabinet as ministers sought funding for their departments' environmental programs. It was in the midst of this politically innovative and transformative period which saw the environment move from the margins to the centre of government attention that IISD was created.

Institutional Structure

IISD is a private, not for profit organization created under the *Canada Corporations Act, Part II: Corporations Without Share Capital*. It is likely that the decision to make the institution a private, not for profit organization was due to the fact that both federal and provincial governments were involved in its creation, making a crown corporation impossible.[6] So, the institution was developed as a kind of hybrid, receiving funding both from the Government of Canada and the Government of Manitoba.

The structure chosen for the organization was a popular model at the time[7] and exemplified in institutions such as the International Development Research Centre, the International Centre for Ocean Development and the Economic Council of Canada.[8] These organizations were characterized by funding arrangements in which the government would provide the funding they needed to define their core business and establish relationships with stakeholders.[9]

The establishment of the institution as an independent organization has been lauded as an "inspired decision"[10] for a number of reasons. First, it has enabled IISD to take risks that would have been prohibitive for government and produce reports that would have been difficult to create in a bureaucratic environment. Second, because it has allowed the institution to appoint its own board members without the intervention of government, allowing IISD to establish a highly qualified Board of Directors, one which has been described as one of the best boards of any like institution. Finally, it has given IISD the freedom to determine its funding structure, giving it a great amount of flexibility to determine how its funds will be used.

Even though it is a fundamentally independent institution, IISD maintains close relationships with its funders. It has entrenched this relationship in the organization's bylaws which grant the president of CIDA, the deputy minister of Environment Canada, and the premier of Manitoba observer status, which allows them to participate in board meetings. IISD has found that the benefit of having major donors participate in this way is that they are able to gain an understanding of what the institution is doing and identify ways in which they are able to collaborate.[11]

In addition to being a Canadian-based institution, IISD is also a fundamentally international organization exemplified by the international representation on its Board of Directors. While board members must be 50 per cent plus one Canadian, this structure has allowed a wide variety of prominent individuals from around the world to contribute to the work of the board. This means that international perspectives are always a consideration for the Board of Directors, an undeniably important characteristic for an organization striving for a voice in international fora.

MAJOR ERAS

During the first two years of its existence, IISD was relatively inactive from an external perspective but internally the groundwork was being laid for the institution. The early days of IISD were spent debating the key internal elements of the organization such as the mission, the structure and the programs. The founding chair, Lloyd McGinnis, a professional engineer and prominent businessman, recalls: "In those early days we spent as much time telling people what we were not going to do as we did outlining our plans. Responding to a question on television in the spring of 1990 in Vancouver, I stated that no, we were not going to spend our funding on bricks and mortar, and no we would not be employing lab coats. As the interview pressure mounted, I somehow blurted out that the Canadian challenge was to convert a concept into practice and we were on our way."[12] Jim MacNeill, former secretary general of the World Commission on Environment and Development, a member of IISD's founding Board and its second chair, also remembers the importance of the early days of the institution: "I vividly remember the Board's first meeting in Vancouver on March 21, 1990. Our task was to provide this new venture with a vision, a mandate and its first few strategic directions. The debate was spirited and continued through several sessions. IISD's position today as one of the pre-eminent institutions of its kind in the world stems directly from those early choices."[13]

By 1991, IISD's mandate had been refined to focus on two main areas: policy research and communications. In addition, research themes had been identified and included the integration of environment and economics in decision-making; institutions for sustainable development; and reforming public policies. Possibly the most important area of public policy research undertaken by IISD was that of trade and sustainable development. This subject faced a considerable amount of skepticism when it was first introduced as a research theme,[14] but it turned out to be a very timely decision. In 1991, to prepare for the Earth Summit the following year, the General Agreement on Tariffs and Trade revived its Group on Environmental Measures and International Trade as a means to address the environmental concerns of GATT member countries.[15] In addition, the North American Free Trade Agreement (NAFTA) which was also negotiated in the early 1990s, was the first major trade agreement which explicitly involved environmental issues. The result was that IISD was at the forefront of research with regards to trade and environment linkages and had developed expertise that allowed it to address the issues faced by both the GATT and NAFTA over the past 17 years.

Communications have always been important for IISD. One key way the institution has changed since its inception is through its use of the internet. As early as 1991, the organization was examining how it could become more connected through information networks. The Annual Report from that year details communications objectives involving the "exploration of international computer networking relationships."[16] By 1993, IISD recognized how much its work had been "materially assisted by the rapidly expanding use and capabilities of microcomputers and information networks."[17] In 1994, IISD launched the organization's website, IISDnet, a fully-electronic database "allowing fast and focused computer access to the Institute's information clearinghouse."[18] This early adoption of internet technology made IISD one of the first non-governmental organizations to have this kind of internet presence. The growth of the internet and electronic communications helped IISD solidify its place as a world leader in sustainable development and an important source of information on environmental issues. A small group of board members are credited for being major advocates of emerging electronic technologies including the internet. The early uptake of this new technology turned out to be one of the most important decisions made by that early Board and has led to IISD's prominence on the internet. This willingness to explore and adopt new forms of information technology is a fundamental element of IISD's success and the focus on the accessibility of reliable and current information about sustainable development has always been a primary concern. By making information accessible and establishing an early presence on the internet, IISD was able to position itself as a leader in information dissemination at a time when most organizations were still relying on traditional methods for getting their data out to their audiences.

The internet is not the only way that IISD has found to expand and improve its commitment to communications. The *Earth Negotiations Bulletin* (*ENB*) is one of IISD's most successful products and began in 1992, at the United Nations Conference on Environment and Development (UNECD), also known as the Earth Summit, in Rio de Janeiro, Brazil. The Earth Summit was a very significant event for the SD community; in addition to the national delegations supporting the 108 heads of state at the formal Summit, a parallel NGO forum drew 17,000 people.

It was in the lead up to this significant event that three individuals – Pamela Chasek, Johannah Bernstein, and Langston James (Kimo) Goree VI – recognized the need for a reliable, timely and neutral source of information to communicate the proceedings of the negotiations to the massive number of delegates. To fill this void, they summarized their notes daily during the first week of negotiations and posted their reports under the name *Earth Summit Bulletin* on NGO networks. The report became an "immediate sensation among government delegates who ended up with copies,"[19] and highlighted the need for this kind of service. During the meetings in Rio, IISD watched this publication with interest and provided support for this new information service. After the Rio Earth Summit, IISD offered the ENB an institutional home. Since then, the service has continued to grow and is now created and distributed at major conferences all over the world as a report on the conference proceedings. In the summer of 1994, the institution launched *Linkages*, a website which publishes electronic versions of ENB and is the internet location for IISD's reporting services. Today, *Linkages* sends out information through list serves to approximately 50,000 members on its electronic distribution list, a figure that is significant due to the fact that this is a by-request-only service. In 2006, the IISD's reporting services covered 56 meetings. This speaks to the importance of this kind of publication and the need for concise and comprehensive reporting of meetings, negotiations and conferences.

A third era in IISD's history is possibly the most substantial in terms of defining how the organization was run. In 1995, the Government of Canada, headed by Prime Minister Jean Chretien, went through the deficit cutting Program Review. At this time, IISD was informed that a representative from the Government would be visiting to notify them of a cut to their core funding. This however, was not simply a cut, it was a "monumental slash"[20] as Environment Canada's funding was reduced by 91 per cent between 1996 and 1998, representing a overall decrease in total core funding of approximately 45 per cent during that two year period.

This funding cut led to a fundamental shift in the way that the institution did business. Jim MacNeill, who was the chair of IISD during this era, had

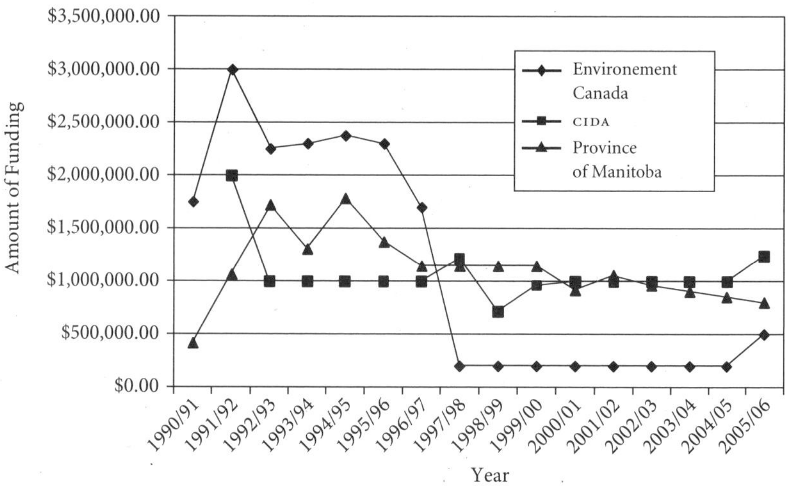

Figure 1
IISD Core Funding 1990–2006
Source: Lillian Hayward based on IISD annual report data

been arguing for the diversification of funding and this funding cut gave IISD the impetus it needed to change from a spending culture to a revenue culture. With that: "A major effort [began] to expand the sources and levels of the Institute's funding. This funding transition mark[ed] a significant change in institutional culture for IISD, with a very successful staff effort to find support for programs. Today, IISD's annual budget is double the level of 1995 expenditures, even though the level of core funding has dropped."[21]

This change in funding sources has shaped the institution in many ways, fixing what has been described as one of the flaws in the original model for IISD. Core funding was wholly given by the government, and its original structure resulted in IISD being totally dependent on this funding with little incentive to pursue creative ideas or activities. By shifting to a bonus system based on budget and fundraising targets, the program directors became fundraisers. This made them responsible for listening to their audiences, shopping around their proposals and raising the funds required to support their projects. This fundamental shift in the orientation of the organization enabled IISD to become a significantly more entrepreneurial organization.

The entrepreneurial culture that emerged from this period is another key factor in the institution's success. In order to survive, IISD had to look outside of itself to determine what its audiences wanted. This resulted in the development of new ideas that were in tune with the international environment and which addressed the wants and needs of its funders. This means that the

Institute does not spend much time responding to requests for proposals from governments and other agencies, favouring instead the development of new ideas and the development of partnerships with donors.[22] One of the major benefits of this model is that it keeps the ideas that are being produced by the Institute new and fresh. It also means that any products that it produces will be taken up by an audience that has already committed to it.

The arrival of William Glanville in 1998 signaled the beginning of a new era for IISD for a number of reasons. First, it represented a strengthening of senior management due to the fact that he was joining as the Institute's first vice-president and became the first chief operating officer (COO). Second, when IISD's president, Arthur Hanson, who was based in Winnipeg, stepped down as president and David Runnalls, based in Ottawa, took over as interim president, and eventually assumed the position permanently, Glanville was the senior person in Winnipeg. This arrangement has allowed IISD to have senior representation both in Ottawa and Winnipeg.

With the addition of a COO to its staff, IISD could begin to examine institutional development and begin to establish a more coherent approach to program planning. The result was a new strategic plan which was presented to the Board of Directors in 1999.[23] The plan was developed to coincide with IISD's tenth anniversary and mark the beginning of a new five-year programming cycle. This new plan had to reflect the new realities of the organization which included substantial growth of its revenues from about $5 million in 1993 to almost $10 million in 2000. Also, the Institute had by this time expanded from its single office headquartered in Winnipeg to include offices in Calgary, New York and Ottawa.

The new plan helped define IISD's vision – "Better living for all – sustainably" – and its mission – "To champion innovation, enabling societies to live sustainably"[24] – both of which are still used to define the Institute. An internal strategic review by the Board of Directors and staff led to a reorganization of IISD away from what had been a rigid program structure and towards a more dynamic configuration in order to "capture the energy of the entire staff to encourage creativity, innovative thinking and interdisciplinary research."[25] This was done by redefining programs as "strategic objectives" and allowing employees to move between these objectives according to where their expertise was needed.[26]

The next big change for the organization will likely come when its current president, David Runnalls, leaves the Institute at the end of 2009. He is the longest serving president in IISD's seventeen year history, having been at the helm of the organization since 1998. A change in this kind of long-standing leadership will undoubtedly mean a significant change for IISD though the ways in which this change will manifest itself will vary depending on who comes into this position.

BARRIERS AND CHALLENGES

IISD's early life was marked by a "series of birthing and budgetary crises and a couple of near-death experiences."[27] These were significant obstacles that the organization had to overcome and, in many cases, have helped shape the organization into what it is today. The first crisis for IISD came only a few years into its life when the Mulroney government terminated several institutions created around the same institutional model as IISD "as part of a wider policy of expenditure reduction."[28] For example, both the Economic Council of Canada and the International Centre for Ocean Development both found themselves on the chopping block. The termination of these and other small organizations like them only saved the government $12 million a year, meaning that these cuts were not necessarily the deficit reduction or streamlining measures the government claimed,[29] implying that the government was no longer interested in supporting organizations that were established according to this model. While there were concerns that IISD could see its end in another round of similar cuts, two critical factors helped secure its survival. The first was that IISD received provincial as well as federal funding. Indeed the Province of Manitoba was extremely loyal to IISD. It had lobbied hard to ensure this crucial sustainable development institution would end up in its capital and did not want to see it taken away. The second factor was that it was Mulroney himself who created the institution, thereby making it very difficult for him to do away with it.

The second crisis came in 1993. "It started with the June 1993 Kim Campbell reorganization, in which [the Department of Environment] suffered significant losses to its mandate, personnel, and budget."[30] This decline in the resources of the department impacted IISD as Environment Canada clawed back some of its funding. While this was by no means a fatal blow for IISD, there was a real concern within the organization that it would set a precedent which would result in Manitoba pulling back its funding too when it saw the federal government failing to fulfill its part of the funding agreement. While this situation never materialized, the potential for it to happen caused a great deal of stress within the organization.[31]

The third crisis was the Liberal's Program Review when, as discussed above, IISD suffered a severe cut to its core funding. The potentially fatal slash to the institution's core budget could have been its demise, however it resulted in a significant shift in the way it did business, moving from an expenditure driven to a revenue-driven organization. The Institute emerged from this crisis as a more entrepreneurial organization which was able to adapt and change according to the desires of its audiences. This outcome has in fact made it stronger by enhancing its ability to respond to the sustainable development community's needs and putting it in a better position to respond to the international community's interests.

Each of these crises presented a challenge for the newly formed institution and could have meant the end before it even reached its fifth anniversary. However, instead of destroying IISD, they actually forged the organization into what it has become today. "Each of these crises inspired the Board, management and staff to new heights of leadership and determination, and from each the Institute emerged stronger and more vigorous than ever."[32] Overcoming the early crises and challenges is a testament to the strength of the institution and a demonstration of its ability to overcome even the most potentially fatal blows. Fortunately, no crises of similar magnitude have materialized in recent years but the organization still continues to face ongoing challenges. From the very beginning, the institution has had to perform a balancing act between the interests of all of its stakeholders. These stakeholders are multiple and varied, from those in its home province of Manitoba, to those at the national level, to the international community. Angela Cropper, IISD board member and international vice-chair, explains:

Finding the right balance between attending to the needs of the home country and addressing the needs of the rest of the world, in keeping with the institute's vision and mission, is a recurring dilemma around the Board table. I have often been found in the posture of holding the institute's feet to this fire. But perhaps this is the role of its International Vice-Chair! And recent recognition that IISD is the most highly ranked and researched sustainable development policy outfit, globally speaking, is a good indicator that it might be successful in managing this dilemma.[33]

In spite of this success, overcoming this challenge is a constant balancing act that the institution must face on an ongoing basis.

In addition to the crises and internal challenges addressed above, IISD has also faced a number of external barriers. Perhaps the most important is through the various "waves of environmental interest" that have seen environmental fervor rise and fall within the Canadian population. The Bruntland Commission's definition of sustainable development was widely accepted and captured the interest of governments, industry and individuals. However, as the concept of SD permeated through the population it came to mean everything to everyone. Also, while environmental issues were "top of mind" in opinion polls in the late 80s and early 90s, they quickly dropped off the Canadian population's radar during the recession of 1992–93 and as a result of other highly publicized political events such as the failures of both the Meech Lake and Charlottetown Accords. This combination of a concept that is difficult to define and a diversion of public interest away from environmental issues created a barrier for IISD's work.

This is closely linked to another struggle face by the institution and that is the challenge of being relevant. At the beginning of its institutional life, IISD had to find ways to become relevant to the audiences that it most wanted to reach. This meant, developing brand recognition, gaining trust and firmly establishing itself as a reliable source on sd. Remaining relevant continues to be a challenge for iisd. As part of its regular programming cycle, the Institute reviews its internal capacity and external issues in an attempt to determine the best fit for its expertise. Its entrepreneurial structure helps address this challenge as program directors strive to "sell" proposals to funders that are both current and relevant. Relevance is something that iisd strives for and something which it is able to attain because of its structure.

OPPORTUNITIES AND BREAKTHROUGHS

It is the barriers and challenges that have shaped the structure of iisd but it is the opportunities which have made the most significant contribution to iisd becoming a world renowned institution. One of the major breakthroughs in iisd's history is their involvement in the 1992 Rio Earth Summit. Not only was it a globally significant event, it was in many ways iisd's debut on the international stage. The Institute could not help but become involved when one of its own board members, Maurice Strong, who had played a significant role in the Brundtland Commission, was named the Conference's secretary general. In addition, iisd "made commitments of both human and financial resources to certain projects contributing to the unecd preparatory process,"[34] and used the event as an opportunity to release its first major report *Business Strategy for Sustainable Development: Leadership and Accountability for the 90s*. Following the Summit, Lloyd McGinnis, chair of iisd's Board of Directors noted:

Our presence was felt in several ways: the contributions of Nicholas Sonntag, our Communications and Partnerships Director, who worked directly with Maurice Strong; the daily publication of the Earth Summit Bulletin; the participation of several Board members and our President with the Canadian unced delegation; co-sponsorship for several events at the Global Forum and a display booth there; and financial support for developing country nationals to attend specific events.[35]

The daily release of *Earth Summit Bulletin* allowed people to become genuinely involved in the meetings and enabled everyone, including governments with small delegations to keep up to date on outcomes of key negotiations, something that had previously been the domain of wealthy countries with large delegations. This meant that communications improved between parties due to its use as a common knowledge base. iisd supported the Bulletin at the Earth Summit and saw the extensive benefits that came from it. The Institute offered the *Earth Summit Bulletin*, renamed the *Earth Negotiations Bulletin*, an institutional home and it has been part of the organization ever since, providing a vital service for un Conferences and Summits. "More than fourteen years later, iisd Reporting Services has produced thousands of reports from hundreds of negotiations covering dozens of major multilateral environmental agreements."[36]

After taking up the ENB, the challenge then became finding a fast, efficient and effective way to distribute it and other IISD publications. At the recommendation of their information technology advisor, IISD became one of the first 1000 users of the World Wide Web.[37] This technology has allowed IISD's information to reach a wide range of users ever since.

IISD program areas also demonstrate areas where the institution has embraced opportunities. Every five years the organization does an external and an internal scan in which it determines what issues are presenting themselves as environmental challenges and what it is possible for the institution to do within its capacity. Intentionally IISD has tried to stay away from opportunities in crowded fields, preferring instead to look for opportunities in areas where it would be possible to create new perspectives.[38] This has led to important work in fields such as trade and sustainable development, an area that IISD has been working in since the very beginning of its mandate and an area which gained prominence when environment and development became major considerations in both NAFTA and the GATT. This expertise in trade and SD has become well respected by institutions such as NAFTA as demonstrated by the fact that in 2004, IISD was given special amicus status by a NAFTA Tribunal in Methanex v. The United States, "making IISD and a U.S. NGO the first civil society groups to be recognized in this way and a major step forward for the transparency of such processes."[39]

IISD has experienced a number of breakthroughs since its creation but many of these are the result of opportunities that it has created for itself. These have been attributed to a strong Board of Directors and Senior Staff who have been able to identify which opportunities to pursue and which risks to take. This has enabled IISD to take advantage of major meetings, new communications products and program areas which foster relationships and establish credibility.

ORGANIZATIONAL ISSUES

Finances

The original funding agreement signed at the Globe 90 Conference, it was for a five year, $25 million deal in which Environment Canada would contribute $3 million, the Canadian International Development Agency would contribute $1 million and the Government of Manitoba would contribute $1 million annually. This was guaranteed core funding to help build the organization as well as establish and maintain relationships.

The problem with this kind of funding model however, is that because all of the organization's most important and immediate costs are covered, there is no incentive to think creatively or seek out audiences for the work that is being

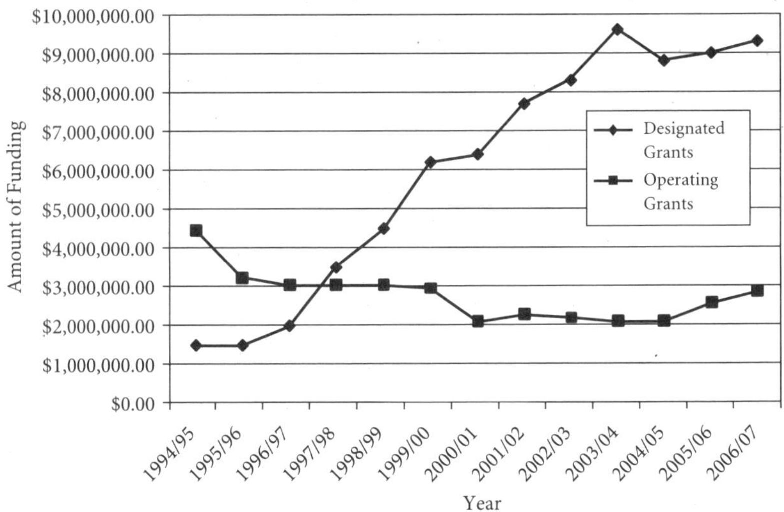

Figure 2
IISD Designated Grants and Operating Grants 1994–2007
Source: Lillian Hayward based on IISD annual report data

done. Cuts by Environment Canada resulted in serious and significant funding reductions for the organization. Rather than allowing these changes to weaken the Institute, IISD adopted a new structure in order to expand the sources and level of its funding. For instance, Jim MacNeill noted at the time of IISD's fifteenth anniversary: "When the government instituted massive cutbacks in the mid '90s, for example, we transformed the institute's culture. It became markedly less dependent on public money and emerged stronger and more vigorous than ever."[40] So, in spite of the fact that Environment Canada cut its core funding from a high of $3 million down to a low of $200,000 a year, the Institute's overall funding has nearly tripled from 1993 levels, from about $5 million then to approximately $14 million today. The majority of this increase has come in the form of designated grant funding, the funding that IISD seeks out itself to fund its programs. Designated grants started outpacing core funding in 1998 and have maintained levels well above core funding ever since.

This demonstrates the strength of the survival instinct of this organization. The serious and significant threat posed by the drastic reduction in funding meant that it had to find other sources or risk perishing. By diversifying its funding sources, IISD has created a funding base that is insulated from drastic changes. Recently, IISD has made another shift in its funding model. While still receiving operating grants and designated grants, it has started introducing what it calls "framework agreements" in which donors commit to providing funding for both core operations and programs over multiple

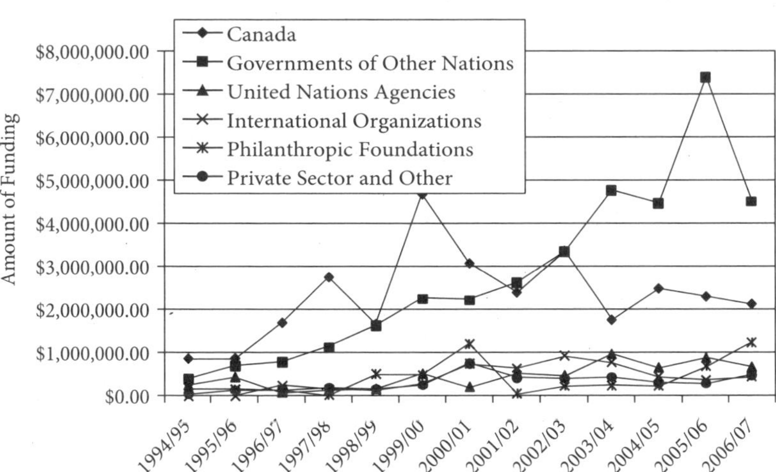

Figure 3
IISD Designated Grants 1994–2007
Source: Lillian Hayward based on IISD annual report data

years. The benefit of this new kind of funding is another way of diversifying its funding arrangements and guarantees funding for a specific period of time. Having organizations enter into these kinds of agreements helps IISD form closer strategic alliances and partnerships with its donors.[41]

Looking at IISD's funding demonstrates the positive reputation that it has developed in the international community. Since 2002, IISD has consistently had more designated grants coming from governments ·outside of Canada than inside, a reflection of the value that is placed on its work by governments around the world. In terms of specific country contributions, Canada remains a top donor, with the Government of Canada contributing approximately $1.54 million in designated grants in the 2006 – 2007 fiscal year. However, the Government of Switzerland is not far behind, contributing a total of approximately $1.33 million in designated funding in 2006–2007.

Personnel

The very first Board of Directors was appointed by the Government of Canada but after that, the Government has had no part to play in appointments to IISD – on the board or staff. The independence of this institution has resulted in the freedom of IISD to appoint its own board members and led to the appointment of very strong Boards of Directors.

Staffing of the institute has evolved into a very flexible and adaptable system. When it was first established, IISD was housed solely in Winnipeg and staffing was done for this office. However, since then the Institute has established permanent staff in offices in New York, Ottawa and Geneva and created "associate" positions around the world. This model was introduced in order to attract highly qualified individuals through open-ended but formalized contracts.[42] This innovative staffing tool provides the flexibility required for these positions while maintaining certain bureaucratic elements that are necessary for this kind of employment arrangement. The result is that IISD has been able to expand its workforce beyond its regular employee base and attract the expertise of highly qualified individuals from all over the world.

With the addition of this associate structure the Institute has also become flexible in terms of work as the well established IT infrastructure allows people to access everything electronically regardless of their physical location. This has allowed IISD to grow its employee base by adding approximately 40 associates to the workforce. A structure like this one offers the ultimate flexibility in terms of finding the right person to do a specific job. This is beneficial because the Institute must compete in the international community for specialized expertise and skills. Capacity has always been a challenge for IISD[43] and attracting the very best people in a crowed international marketplace is difficult. This kind of hiring arrangement offers the flexibility of both time and location which helps attract highly qualified individuals, allowing the organization to hire the best and the brightest at the moment that they are the best and the brightest.

Having access to this kind of capacity is extremely important for an institution that is continuously striving to keep itself relevant. In addition to this, having associates enables IISD to reach beyond its organizational boundaries to strengthen networks through individuals located all around the world. Another benefit is the ultimate flexibility both for those who are employed as part of this arrangement as well as IISD. By giving its staff the opportunity to work anywhere and their associates the chance to work in a flexible system, IISD is able to attract and retain some of the best minds in SD. This structure helps to create a nimble and adaptable institution that is able to thrive in a competitive international environment.

Location

An interesting feature of IISD is the fact that it is based in Winnipeg, Manitoba. To understand one of the key reasons that Winnipeg was selected as the home of this organization, it is necessary to look back at some of the events that occurred in the Canadian aerospace industry in 1986.[44] At that time, the Government of Canada was offering a $1.4 billion contract for maintenance of CF-18 fighter jets. Following bidding and review by a seventy-five member panel at the Department of National Defence, Bristol Aerospace Limited of Winnipeg was the top choice, [45] both underbidding Montreal's Canadair, by $3.5 million and receiving a substantially higher technical rating.[46] However, the company was passed over in favour of Canadair, stirring up considerable anger in Western Canada. In response to the announcement Winnipeg MP Lloyd Axworthy, a member of the opposition, stated: "It's a clear message to Western Canadians that we should be hewers of wood and drawers of water … We're not capable of undertaking major activities in technology development. It's an unfair and tragic message – one that has to be fought against."[47]

Shortly thereafter, when the federal government's intent to create an international institute for sustainable development came to light, the Province of Manitoba was quick to express its interest. As Gary Filmon, former premier of Manitoba explains:

When my senior staff and I heard about this in a news clip the next morning, we wondered what the mandate was and where it would be located. Since we had just established the Manitoba Round Table on Environment and Economy in agreement with Canada's response to the Brundtland Report and I was the Chair, I immediately wrote to the Prime Minister to propose that this new institute should be located in Winnipeg in recognition of our commitment to the Round Table process and sustainable development. We were told there were a number of locations under consideration, including Montreal.[48]

This initial proposal was followed by intensive lobbying of Environment Canada and the Prime Minister's Office by the Government of Manitoba. The lobbying exercise was successful and led to a subsequent funding agreement which created IISD.

Some have criticized IISD for its location, claiming that its growth is inhibited because it is so far away from the major urban and economic centres of Vancouver, Toronto, and Montreal as well as Canada's centre of government in Ottawa. However, like many of the challenges that IISD has faced in the past, the institution has turned this into an opportunity, touting what it refers to as "the Winnipeg Advantage." "[B]eing in Winnipeg afforded IISD greater access to local decision makers and allowed its messages to be heard locally and provincially, not drowned out by the background noise of national headlines in larger centres."[49]

Since its earliest days, IISD has stressed the importance of communication. Finding new and innovative ways to communicate was not only necessitated by its mandate but also by its location which encouraged the institution to take risks on new technologies like electronic networks and the internet. This early integration of technology has led to the recognition that in today's world of information technology, location is becoming less and less relevant and that it is the work that is produced that actually matters. With the introduction of associates, IISD has been able to adopt a dispersed office model while building partnerships and networks around the world. This model has helped IISD attract individuals regardless of location and has enhanced its ability to compete for employees in an international market place.

Another advantage is the involvement of the Government of Manitoba in the organization. Both premiers who have led the province since the establishment of IISD have been very supportive, as has the City of Winnipeg. Neither Manitoba nor Winnipeg want to see IISD fail and this has helped get the institute through difficult times, especially those faced at the beginning of its mandate. Had IISD been located in a big city like Toronto or Montreal, it is unlikely it would have seen the same level of support during difficult times.

Winnipeg is not a major urban centre and it is not likely the first place people consider when selecting potential locations for a world class institution however, it is this home that has had a large part to play in building and shaping this organization. "Being located in Winnipeg offered an advantage ... because it forced IISD to develop in a way that allowed it to connect to and exert influence on the outside world."[50] This ability, and necessity, to make connections and build networks contributes to the strength of the organization.

THE ROAD AHEAD

IISD has been a very successful institution. As of 31 March 2007, it had more than seventy donor organizations including federal departments and provincial governments in Canada, governments of other nations, United Nations agencies, international organizations, philanthropic foundations and private sector institutions. While much support has come to the organization in the form of funding, it has also come in the form of accolades. In 2004, IISD was declared the Most Effective SD Research Organization in a GlobeScan survey. The survey asked "experts who have either had a direct role in SD research organizations, had dealings with them, or studied them ... to name a maximum of four specific SD research organizations that they consider to be particularly effective."[51] IISD was considered by experts to be more effective than other well known SD institutions such as the World Business Council for Sustainable Development, the World Resources Institute and the United Nations. While comparing these institutions against one another is difficult due to their vastly different mandates and activities, it speaks volumes about the effectiveness of IISD, a small organization in the midst of these large establishments.

In many ways, IISD has exceeded expectations by becoming a world leader in SD information and ideas. However, this does not mean that it can become complacent. If the organization is to remain successful it must continue to build on its strengths while maintaining the flexibility and entrepreneurial spirit that it has been so successful at adapting to its work.

In spite of the successes of IISD as a sustainable development institution in Canada, the organization is an exception rather than a rule. Sustainable development as a concept is alive and well in Canada however, as an agenda for action and change, sd has struggled. In a report released in 2005, entitled *Its Time to Walk the Talk*, the Standing Committee on Energy, Environment and Natural Resources noted that governments and corporations in Canada talk a lot about sustainable development but hesitate to take any real meaningful action on this concept.[52] The notion of sustainable development has permeated Canadian government and industry but it is still often just lip service, failing to constitute real, concrete action for the majority of these organizations.

This has not gone without notice and Canada has faced criticism from both inside and outside of the country. For instance, The Conference Board of Canada has described Canada as a "middle-of-the-road" performer, lagging behind top oecd countries in a number of environmental indicators such as greenhouse gas emissions and hazardous waste production.[53] The Pembina Institute has criticized Canada's attempts to integrate sd principles into legislation and calling them "limited and almost entirely symbolic."[54] Both the World Economic Forum and the oecd have also been critical pointing out that in Canada, very little progress has been made in advancing the principles of sustainable development.[55] Also, at a recent conference called *Facing Forward – Looking Back: Charting Sustainable Development in Canada 1987–2007–2027*, which brought together a significant cross section of sustainable development expertise within Canada and abroad, the feeling was that not enough has been done. One of the most critical and most regularly repeated quotes that emerged during the two day meeting is that Canada's action has been more a kin to "fiddling at the edges of the problem"[56] rather than making any substantial advancement in the field.

CONCLUSION

To sum up, Canada was once a world leader on the international environmental stage but its reputation has been steadily slipping both within and outside of the country due to its failure to actually integrate sd into policy and day-to-day operations. Overall, Canada's engagement with sustainable development has been underwhelming. While the country has produced a world-class institution that is producing relevant information on sd, aside form iisd's work, there has been a serious lack of science/policy engagement.[57] In spite of having strong institutions within the country like iisd, it seems that there has been very little movement of environmental issues and very little uptake of sustainable development. This comes in the face of the dire warnings issued by the Millennium Ecosystem Assessment in 2005, the shocking predictions of the Stern Report in 2006, and the seemingly daily reports of natural disasters and evidence of accelerating climate change. There is a serious lack of urgency in what Canada has done to date and with action more akin to "fiddling at the edges" of the issues rather than striving for real and significant solutions.

Canadians can be very proud of the work that IISD has done in spite of this environment and can hold it up as evidence that an institution that suffered so many "near death experiences" has the potential to become larger and more successful than anyone could have anticipated. But, one institution is not enough. SD needs to be become part of the policy paradigm in all institutions and it is only then that we will begin to see the implementation of a real agenda for change.

NOTES

1 External Affairs Canada, *Address by the Right Honourable Brian Mulroney, Prime Minister of Canada, Before the UN General Assembly* (29 September 1988), 8.

2 "Environment conference weighs solutions," *The Globe and Mail*, 20 March 1990, A5.

3 Canada, *Speech from the Throne* (34th Parliament, 2nd session, v. 131). Available at: www2.parl.gc.ca/Parlinfo/Documents/ThroneSpeech/34–02–e.pdf.

4 Toby Heaps, "Brian Mulroney: No Regrets," Corporate Knights (Summer 2005), 31–2.

5 Robert J.P. Gale, "Canada's Green Plan," *A Study of the Development of a National Environmental Plan* (1997), 112–13. Available at: www.ies.unsw.edu.au/about/staff/robertsFiles/greenplan.pdf.

6 David Runnalls, personal interview, 24 July 2007.

7 Robert Slater, personal interview, 21 September 2007.

8 Arthur Hanson, personal interview, 9 November 2007.

9 Slater, 2007.

10 Jim MacNeill, personal interview, 26 October 2007

11 Runnalls, 2007.

12 IISD, *Sustaining Excellence for 15 Years. 2004–2005 Annual Report* (IISD, 2005), 11.

13 Ibid.,10.

14 International Institute for Sustainable Development (IISD), "IISD Timeline." Available at: www.iisd.org/about/timeline.asp.

15 Bruce Doern and Thomas Conway, *The Greening of Canada: Federal Institutions and Decisions* (University of Toronto Press, 1994), 144.

16 IISD, *Annual Report 1991–1992* (IISD, 1992), 12.

17 IISD, *Annual Report 1993–1994* (IISD, 1994), 2.

18 IISD, *Annual Report 1994–1995* (IISD, 1995), 11.

19 IISD, *Annual Report 2001–2002* (IISD, 2002), 4.

20 MacNeill, 2007

21 IISD, "IISD Timeline."

22 Runnalls, 2007.

23 IISD, *Annual Report 1998–1999* (IISD, 1999), 2.

24 Ibid., 5.

25 IISD, *Annual Report 1999–2000* (IISD, 2000), 4.

26 William Glanville, personal interview, 15 November 2007.

27 Jim MacNeill, "Chair's Message," *Annual Report 1998–1999* (IISD, 1999), 3.

28 Canadian International Development Agency (CIDA), "CIDA's Strategy for Ocean Management and Development, The Lessons of ICOD." Available at: www.acdi-cida.gc.ca/CIDAWEB/acdicida.nsf/En/NAT-329142438–QRX.

29 "Federal axing called political Ottawa chopped Economic Council as part of '92 budget," *The Globe and Mail*, 4 March 1993, B2.

30 Glen Toner, "Environment Canada's Continuing Roller Coaster Ride," in Gene Swimmer, ed. *How Ottawa Spends 1996–97: Life Under the Knife* (Carleton University Press, 1996), 99.

31 Hanson, 2007.

32 MacNeill, "Chair's Message," 3.

33 Angela Cropper, "Anniversary Reflections," *Annual Report 2004–2005* (IISD, 2005), 10.

34 IISD, *Annual Report 1991–1992*, 13.

35 Lloyd McGinnis, "Chairman's Report," *Annual Report 1992–1993* (IISD, 1993), 2.

36 International Institute for Sustainable Development, "Products," IISD Linkages. Available at: www.iisd.ca/about/about.htm#history.

37 Hanson, 2007.

38 Ibid.

39 IISD, "IISD Timeline."

40 MacNeill, "Anniversary Reflections," 10.

41 Glanville, 2007.

42 Runnalls, 2007.

43 Marlene Roy, personal interview, 5 November 2007.

44 Slater, 2007.

45 "CF-18 deal likely to allay Quebec complaints," *The Globe and Mail*, 31 October 1986, A1.

46 "Canadair gets $1.4–billion job jet repair contract stirs bitterness," *The Globe and Mail*, 1 November 1986, A1.

47 Ibid.

48 Gary Filmon, "Anniversary Reflections," *Annual Report 2004–2005* (IISD, 2005), 11–12.

49 William Glanville, *The Winnipeg Advantage: An Example of Organizational Adaptation and Innovation* (February 2005), 1.

50 Ibid.,5.

51 GlobeScan, *The GlobeScan Survey of Sustainability Experts*. Available at: http://surveys.globescan.com/sdroleaders/sose04–2_resorg.pdf.

52 Senate Standing Senate Committee on Energy, Environment and Natural Resources, *Sustainable Development: It's Time to Walk the Talk* (2nd interim report, June 2005), 1.

53 Conference Board of Canada, *Performance and Potential 2004 – 2005: How Can Canada Prosper in Tomorrow's World?* (2004), 40.

54 Pembina Institute as cited in Glen Toner and Carey Frey, "Governance for Sustainable Development: Next Stage Institutional and Policy Innovations," in G. Bruce Doern, ed. *How Ottawa Spends 2004–2005* (McGill-Queen University Press, 2005), 201.

55 Senate Standing Senate Committee on Energy, Environment and Natural Resources, *Sustainable Development: It's Time to Walk the Talk*, 1.

56 Elizabeth Dowdeswell, "Looking Back: How Have We Done in Canada?" Remarks at the conference, *Facing Forward – Looking Back: Charting Sustainable Development in Canada 1987–2007–2027*, 18 October 2007.

57 Alan Nymark, "Looking Back: How Have We Done in Canada?" Remarks at the conference, *Facing Forward – Looking Back: Charting Sustainable Development in Canada 1987–2007–2027*, 18 October 2007.

11 Advocate or Auditor?
The Conflicted Role of the Commissioner of the Environment and Sustainable Development

LAURA SMALLWOOD

Sustainable development (SD) requires a fundamental change in the values and operation of government and society. Change is often difficult, even when necessary. The Canadian government has encountered significant challenges in its effort to mainstream SD into government practice as other issues such as national unity, the fiscal deficit, and health care have often distracted parliamentarians from this long-term issue. Progress on SD requires commitment and adaptation from the business community, all levels of governments, and society as a whole; "sustainable development cannot simply be "delivered" by politicians and officials, but demands an active and creative input from all sectors of society."[1] Nonetheless, the federal government remains an essential actor for achieving SD in Canada. Through their policies, programs, and regulations, and the billions of dollars they spend each year, federal departments have an incredible amount of influence on Canadian society. Thus, government must play a central role in the transition towards sustainability; as leader, coordinator, and through the establishment of appropriate policies, regulations and infrastructure.

In the two decades following the release of *Our Common Future* (OCF) the federal government has made many promises with respect to the integration of SD into Canadian society; however it has failed to deliver on many of them. The pervasive gap between rhetoric and action was a primary driver for the creation of the Commissioner of the Environment and Sustainable Development (CESD). Located within the Office of the Auditor General (OAG), the CESD helps ensure accountability within government for progress on SD. In its first twelve years the work of the CESD has triggered an interesting debate on its design and potential. This chapter will examine its role, the tools it uses to achieve its objectives, key institutional and personnel changes over its early life, the root of the contentious debate it has incited, and its achievements to date. It is imperative to examine the CESD from this historical perspective in order to gain insight into how the CESD has influenced the federal government's performance on SD and what its potential role could be. The following analysis is based on interviews with current and former senior public servants and parliamentarians and through a comprehensive literature review on SD governance in Canada.

THE CREATION: DRIVERS AND POLITICAL LANDSCAPE

The political culture of the late 1980s and early 1990s was greatly influenced by the 1987 publication of *Our Common Future* and the idea of SD. Canada's integral involvement in the development of OCF[2] and executive-level engagement with SD was internationally acknowledged. Indeed, the U.S. President's Panel consulted with Canadian officials to determine how the U.S. could best engage on SD.[3] OCF called on countries to make central agencies and major departments in charge of economic and social policies responsible for SD, as they play key roles in national decision-making that determine whether environmental resources are maintained or degraded. At that point in time, few economic and social departments had a mandate to ensure that the ecological resources upon which their goals depended were sustained for the future. This problem triggered many countries to develop national sustainability strategies.

In 1989, a coalition of aboriginal groups and environmental non-governmental organizations (ENGOs) presented *A Greenprint for Canada* to then Environment Minister Lucien Bouchard. This document contained an ENGO vision of an SD strategy for Canada, including a key recommendation for the creation of an auditor general (AG) for the environment.[4] Specifically, it called for a new OAG-like body, separate and focused solely on environmental issues. This recommendation was complemented by a Private Members Bill brought forward by Liberal MP Marlene Catterall proposing the creation of an environment AG. When asked to comment on these recommendations, then auditor general, Denis Desautel, stated that he didn't think there was a need for an environment AG as the OAG was already performing environment audits.[5]

Between the publication of *Our Common Future* and the historic 1992 Earth Summit in Rio de Janeiro, the Conservative government under Prime Minister Brian Mulroney launched the 1990 *Green Plan for Canada*, which did not include the idea of an environmental AG. The *Green Plan* was one of the very first national SD strategies in the industrial world.[6] While dedicating large expenditure programs to cleaning up past mistakes, the SD framework of the *Green Plan* was based on the need to change societal decision-making systems to prevent environmental damage in the first place, while sharing responsibility with business, provincial governments, and individual citizens. This forward-thinking "anticipate and prevent" approach was much different from the traditional "react and cure" orientation of environmental policy. Included in the *Green Plan* was a statement that federal departments and agencies would implement enhanced policies and procedures for environmental auditing in order to evaluate organizational compliance with environmental standards and policies and measure performance against desired goals and objectives.[7] Despite the progressive initiatives proposed, both the Liberal opposition and many environmental groups argued that the *Green Plan* did not go far enough to secure a sustainable future for Canada.

The Earth Summit increased pressure on Canada to incorporate SD into its governance system. As a result, Environment Canada (EC) held a meeting following the Summit with 25 individuals from within the OAG and EC, as well as other stakeholders, to discuss the idea of including an environmental auditor general within the OAG. The participants from EC were surprised to learn of the audit work already being done on the environment in the OAG. Although the meeting began on a territorial and competitive note (EC vs. OAG), it ended with an acceptance that the OAG had certain competencies in the environment realm.[8] This understanding would influence the decision process with respect to the CESD a few years later when the 1993 election ended the Conservatives" ten year reign.

During the election campaign all major political parties committed themselves to advancing the integration of SD. Being highly critical of the *Green Plan*, the Liberals made specific promises to further the implementation of SD.[9] In fact, chapter four of the Liberal election platform *Creating Opportunity: The Liberal Plan for Canada* was dedicated entirely to environment and SD and was authored by then opposition environment critic Paul Martin. It called for an environmental auditor general that would report directly to Parliament and hold individual departments and the government as a whole accountable for how successfully federal programs and spending were supporting the shift to SD.[10]

THE GREAT DEBATE

Implementing the promises of the Red Book was the principal occupation of the Chrétien government during its first few years in power. In March 1994, the Minister of the Environment Sheila Copps asked both the House of Commons Standing Committee on Environment and Sustainable Development (SCESD)[11] and a team at EC to consider a number of functions related to the campaign promises and determine how they could best be performed, including the establishment of an environmental auditor general who would ensure that the government's actions were carried out in a sustainable manner. The challenge was to define the mandate for such a position and find a home for it;

however, the parliamentarians and bureaucrats came out with opposing views and recommendations on both the position and the mandate.

Environment Canada's Proposal

The group at EC proposed the creation of an environmental auditor general within the OAG. Some of the strongest arguments for this structure were that the OAG already had an excellent reputation for unbiased and thorough audit work. The OAG also had credibility with the public, departments and politicians, which EC argued would make the audits appear more reputable. As well, EC argued that the OAG already had an established environmental auditing group with the required training, methodology, documentation and reporting structures. Having a separate body outside the OAG was therefore thought to be unnecessarily duplicative.[12]

Standing Committee on the Environment and Sustainable Development's Proposal

The SCESD concluded that the most appropriate way to deliver on the Liberal government's proposed functions was through the creation of a Commissioner of the Environment and Sustainable Development (CESD), as well as an expanded role for the OAG to enhance the accountability of existing public policy related to the environment. The SCESD was chaired by the Trudeau-era Environment Minister Charles Caccia. According to Brian Emmett, the first CESD, the fact that the SCESD was chaired by "a real environmentalist" was extremely influential in the recommendations proposed. The approach called for was to mirror that taken for other vital issues to Canada, such as human rights and official languages. Although the terms of reference provided by Copps stated that "the Committee should bear in mind the Government's commitment to budgetary restraint and consider the implications of a new office on resources and efficiency," the Committee recommended what they felt would be necessary to put Canada on a sustainable path, given that the independent policy review/evaluation function was the biggest gap in the government's SD infrastructure.[13]

The Committee recommended a separate agency and legislation for the commissioner so that the office would be free to comment on and provide advice on policy, which an auditor is not. The enhanced environmental auditing function in the OAG would then compliment this policy and advocacy role by providing insights on past performance and holding departments accountable. The SCESD report contained many statements from stakeholders about the perils of a purely audit mandate for the role, which foreshadowed the experience that would follow a decade down the line:

If we had to limit the role of the commissioner or auditor to a simple audit, his mandate would be too restrictive for his work to have any real impact on government actions. There would be no possibility for the public to ensure that there is a better analysis of the extent to *which the federal government can respect its mandate to protect the environment. – Yves Corriveau, then director of the Centre Québécois du droite de l'environnement*[14]

The function of an environmental auditor general is at the wrong end of the process [because] the audit of how well existing policies are implemented does little to inform the need for new policies. – *Art Hansen*[15]

It would be preferable to have no commissioner than to have a commissioner whose terms of reference are so restrictive that she or he cannot influence the substance of policy implementation and content and interpretation of laws on environmental issues. – *Kenny Blacksmith, deputy grand chief, Grand Council of the Cree*[16]

However, EC had arguments against an independent commissioner, including the point that an unelected parliamentary advisor was inappropriate and that the minister of the environment should be the one guiding policy as an elected member of cabinet responsible in the House of Commons. They felt that a commissioner would be a potential competitor for EC and that adding another person focussed on policy was not going to improve the Canadian system. The Department of Finance was also concerned about the implications that a commissioner would have, as SD is an economic issue as well and the department wanted to maintain its monopoly on priority setting. As former Commissioner Brian Emmett stated, what unfolded was a "recipe for unhappiness" as the structure that emerged disappointed most in the SD community.[17]

By April 1995, the government responded to the recommendations of EC and the SCESD with the decision to in fact create a commissioner of the Environment and Sustainable Development, but to limit its mandate by housing it within the OAG. The mandate was almost exactly that proposed by EC. Copps later stated that virulent bureaucratic opposition to an independent commissioner had influenced the final outcome.[18] The fact that the approach chosen was less expensive and did not necessitate a new bureaucracy was likely a significant consideration in a time of program review and budget cuts.

The implications of having a commissioner within the OAG were not well thought out as, according to those who were present at the time, not much consideration was put into the "commissioner" title. The title was chosen to demonstrate that environment was special and avoided setting up a separate office for an environment auditor general. However, the confusion this nomenclature would create would prove to be highly problematic in the future.

INSTITUTIONAL STRUCTURE

The uniquely Canadian approach decided on was to be driven from the bottom up via departmental sustainable development strategies (sdss) that would allow sd to be tailored and relevant to each department's unique situation. Both Canada's novel approach to implementing sd in government and the cesd as an institution were eagerly watched by other nations:

What makes the federal government unique among oecd countries in terms of sd governance has been its choice of a decentralized approach to sd planning involving a legislated obligation on most federal departments to prepare their own sd strategies every three years and the establishment of a new institution – cesd – to audit performance and report to Parliament. The premise behind this approach is that these complementary requirements will establish a self-reinforcing cycle of continuous improvement.[19]

The cesd was provided with \$3.5 million to get the office established and to hire 40 staff over its first three years. Since its inception in 1878, the oag's focus had traditionally been on financial audits, however in the 1970s its mandate was expanded to include the "three Es" – economy, efficiency, and effectiveness. The 1995 amendments to the *Auditor General Act* formally added a fourth "E"; from then on the "environment" was to be included among the considerations the oag would have to take into account when determining what to report to the House of Commons. The cesd was assigned two primary roles: reviewing how well government policies, programs and spending support Canada's move towards sd; and providing liaison, monitoring, and encouragement to government, parliamentarians, and the public on sd.[20] In June 2007, the passing of Bill C-288 gave the Commissioner additional responsibilities to include Canada's progress on implementing its climate change plans and meeting its obligations under the Kyoto Protocol in its annual report to the House of Commons.[21]

The ag Act confers obligations on most federal departments. Ministers are given the responsibility to prepare departmental sd strategies and are required through the petitions process to respond to petitions received by the oag within 120 days.[22] The commissioner monitors both the outcomes of the departmental sdss and the ministerial responses to petitions, reporting annually on them to the House of Commons. The Act grants the federal government the power to create regulations describing what should be in the departmental strategies. Appointed by the ag, the cesd is an assistant auditor general who leads a group of auditors and other professionals specialized in environment and sd. The relationship between the cesd and the ag is primarily an administrative one, and as such all documents and statements coming out of the commissioner's office have to be approved by the ag before being released to the public or Parliament. In general, the CESD has room to influence the direction of the office of the CESD (within its mandate) and has control over finances. The AG Act provides a fairly broad mandate for the CESD, stating that he/she can report on anything that is believed should be brought to the attention of the House in relation to environmental and other aspects of SD. Given this discretion, the personality and vision of the CESD can have a large influence on the outcomes of the office of the CESD.

As a result of being within the OAG, the CESD is restricted from performing any policy guidance or advocacy. The lack of the advocacy and policy support was intended to be made up for by government leadership; the CESD was to be only one part of an overall approach to SD.[23] Initially, the Liberal government tried to help fill in this gap through the *Guide to Green Government*, its main commitment to SD and the first government corporate view of the SD challenge. The document stated: "We want to play a leadership role in turning sustainable development thinking into action. This is why we are now taking the next step of establishing a framework in which environmental and economic signals point the same way; a framework which integrates sustainable development into the workings of the federal government – right across the board."[24] Signed by all ministers and the PM, the *Guide* was symbolic of a united government commitment to the principles of SD and the perception of a shared responsibility between parliamentarians and the executive to help transition Canada towards a sustainable future. However, the broad framework provided in the *Guide* was an insufficient proxy for an elaborated federal strategy and failed to provide the guidance required by federal departments to make the SDS process a success.

Supportive Institutions

Environment Canada (EC) has played an important role in establishing and guiding the SDS process. On behalf of the federal government, EC coordinated the tabling of the first round of SDSs, chaired the Sustainable Development Coordinating Committee of Deputy Ministers, and currently chairs the Interdepartmental Network for Sustainable Development Strategies whose purpose is to share experiences and good practices across the federal government. EC has also contributed to the effort to help the struggling SDS process by twice providing supplementary guidance documents for producing the strategies. However, the efforts of EC have not made up for the lack of guidance and leadership from the center of government. As well, EC's continual leadership role in this process is reinforcing the misconception of SD as a purely environmental issue.

The SCESD is a very active parliamentary committee that holds hearings and produces reports on SD and environment issues. In the December 2007 Hill Times Poll of Parliamentarians, the SCESD was considered to be the best

House committee on Parliament Hill.[25] As the CESD's mandate is to serve parliamentarians, the SCESD assists the CESD informally by providing a sense of what issues politicians are interested in. At its best the SCESD highlights the content of the CESD reports through hearings, provides engagement on audit issues, and chastises non-performers. However, "while it has managed to expand the debate in several areas, it has seldom been able to convince Cabinet to adopt its recommendations."[26] While departments go through the SDS process and the SCESD hearings hold non-performers to account, at the end of the day departments continue to do what their ministers direct them to.[27] The Senate Committee on Energy, Environment and Natural Resources" is less active but has also contributed to the current dialog on SD in the federal government. Its Report, entitled *Sustainable Development: It's Time to Walk the Talk*, demonstrates a continued willingness to deal with the implementation challenge that plagues Canada's efforts towards SD.[28]

Other commissioners, within Canada and without, have had an important influence on past and current debates around the CESD's institutional structure. In Canada, the Commissioners of Official Languages and Human Rights have provided examples of independent commissioners who act as an advocate and provide policy support on issues of importance to Canadians. Although their structures and mandates are markedly different from the CESD, they provide an interesting comparison and demonstrate how independent advocates can operate within the Canadian federal system. Commissioners of the Environment in New Zealand and Ontario have also been helpful in identifying useful models. The recommendations made by the SCESD in 1994 were greatly influenced by the examination of the experience of these two offices, as have been the recent debates occurring within the House of Commons.

TOOLS OF THE COMMISSIONER

In addition to the formal audit and reporting tools detailed below, the CESD performs numerous speaking engagements which raise awareness of the importance of environment and SD for Canada.

Public Petitions Process

The public petitions process fulfills the ombudsman role of the CESD and provides a mechanism for citizen participation. Through this process Canadian citizens can demand that government look into issues related to environment and SD in a timely manner. It can also be used to ask federal ministers to explain federal policy, investigate an environmental problem, or examine the enforcement of environmental legislation. Once a petition is received, the CESD must send it to the minister(s) of the responsible department(s) within

15 days. The minister then has 120 days to respond. Since the process began in 1996 over 250 petitions have been received and over 450 responses have been issued. Although the number of petitions received is increasing, from about five petitions a year in 2001 to around 30 a year in 2007, the process is still greatly underused.[29]

A full review of the petitions process was conducted as part of the Commissioner's 2007 annual report to the House of Commons. Users of the petitions process recommended that the CESD take a more active and prescriptive role in responses to petitions by compelling action by federal departments on environmental problems. However, that role – although ideal – goes beyond the authority granted by the *Auditor General Act*. The CESD can provide guidance on preparing and responding to petitions, but cannot determine the content of petitions or compel a particular response or action by a department. The lack of ability to support or compel departmental action when necessary, limits the effectiveness of the petitions process to effect real change.

Performance Audits Applied to Environmental Issues

The auditing process is an essential element of environmental governance as it provides citizens with an independent view of value for money and government performance. The audit capacity within the CESD is widely considered to be one of the institution's greatest strengths and has garnered Canada the reputation as a world leader in environmental auditing. Although it was initially challenging to integrate the performance auditing of the CESD with the financial auditing of the OAG, AG Sheila Fraser created a "five-year one-pass plan" to integrate the performance audit with the rest of the CESD's audit resources.[30] The CESD performs mostly performance audits, with limited financial assessments/analysis in the chapters of the commissioner's Reports. The performance audits of the CESD attempt to answer the following two questions: are programs being run with due regard for economy, efficiency, and environmental impact? and, does the government have the means in place to measure their effectiveness?[31] Financial assessments of SD in federal departments have proven to be very difficult due to what has be characterized as "poor departmental accountability systems," along with the horizontal nature of the issue.[32]

Sustainable Development Strategies

The SDSs are the CESD's primary tool for ensuring that federal departments are incorporating SD into their plans and priorities. These strategies are the lynch-pin of the Canadian bottom-up approach to SD, providing an opportunity for individual departments to uniquely integrate SD into the fabric of their organizations in a manner tailored to their needs:

federal departments and agencies have a significant influence on just about every aspect of Canadian society. The sustainable development strategies and the process of creating them were intended to ensure that before deciding on their policies and programs, government departments would consider the potential consequences – social, economic, and environmental. The strategies were to ensure that departments and agencies understood their opportunities for sustainable development and addressed them in concrete action plans.[33]

Certain federal departments are legislated to create an SDS in which they must elaborate policy objectives, action plans and benchmarks for measuring progress on SD. The strategies must be tabled in the House of Commons and updated every three years, with each department posting its strategy on its official website. Each department reports annually on its activities in pursuit of their commitments made. The commissioner also performs audits on certain departments when deemed necessary, the results of which are published in the annual Commissioner's Report. The publishing of results is intended to encourage departments to adhere to their commitments to avoid public embarrassment.

Although the strategies have done well in raising awareness of SD within departments and have increased the departments" capacity to analyse the environmental dimensions of their programs, they have not achieved their overall goal of making departments and broader Canadian society more sustainable. The CESD has been highly critical of most of the SDSs produced to date and has made numerous recommendations for improvement. Despite these recommendations, little progress has been made. Consensus, both inside and outside of government, has been that the strategies are a bureaucratic paper-pushing exercise and are not achieving their intended goal. The fourth round of strategies are now completed and little has changed: "The ambition and momentum that existed in the early stages of the government's sustainable development strategy initiative has faded. In our view, the preparation and tabling of the strategies have become little more than a mechanical exercise, required to fulfill a statutory obligation. Departments may be meeting the letter of the law with their strategies but most are certainly not responding to the spirit of it."[34]

It can be difficult to perceive Canada as being a leader on SD when its principal tool of integration is habitually characterized as a "compendium of business as usual," bereft of the innovative ideas on how to improve SD across Canada that they were originally intended to produce. The strategies are also not delivering on the long-term change that is necessary for SD. The nature of three year renewal cycle appears to have resulted in a tendency toward short-term goals and a focus on today's issues rather than tomorrow's.[35] Perpetuating the problems within this process, junior or mid-level analysts and consultants are typically given the task to produce the SDSs – which often have no quantifiable targets or timelines.[36] In the 2007 AG report to the House of Commons AG Sheila Fraser stated, "Senior managers ... have not demonstrated that they take the strategies seriously, and few parliamentary committees have considered them."[37] Although often those creating the strategies may be passionate about the process, the disengagement from the decision-making center has created a barrier to innovative initiatives making their way into the broader departmental plans and priorities.[38]

The strategies are also challenged with respect to coherence across departments and the lack of co-ordinated efforts among departments, sectors and levels of government. They remain an inward-looking exercise, focussing on internal capacity building and greening operations rather than activities directed at influencing the behaviour of Canadians. This is especially problematic as it is not the department's physical infrastructure that creates a large ecological footprint but rather their policies and programs that encourage or enable unsustainable activities. Overall, the commitments made in the strategies tend to be modest and vague and are accompanied by soft policy tools (such as information) rather than the hard tools (like regulation) that are necessary for fundamental change. As departments know their performance on these commitments will be audited and reported publicly, there are perverse incentives inherent in the strategies to make modest commitments. The many problems with the SDS process and few benefits suggest that if these strategies were not entrenched in legislation they most likely would have been fundamentally revised or discontinued years ago.

Studies

Although informally discontinued in 2002, studies were originally included as chapters in the CESD annual reports. Their purpose was to help generate thinking around SD within government and to provide parliamentarians and public servants with best practices to help them in their efforts to integrate SD into government operations. The studies were often highly exploratory and varied greatly in their objectives[39]; focussing primarily on the knowledge gap and capacity building, but also included a study on the social aspects of SD.

Through their potential policy implications, studies made the CESD even more different within the OAG. They were considered by many within the OAG to be "audits that are not strong enough," as they do not have the same rigour or methodology as an audit. The OAG perception that they were too policy-driven resulted in their eventual demise.[40]

The Commissioner's Annual Report

The CESD Annual Report is presented to the House of Commons and details in separate chapters the extent to which departments have implemented their

SDSs and the results of audits performed on environment and SD programs of departments. "The reviews of departmental strategies and government programs have been hard-nosed and convincing, earning the Commissioner's office a solid reputation for high-quality and authoritative work."[41] The Reports assist parliamentarians in their oversight of the federal government's efforts to protect the environment and foster SD by providing them with objective, independent analysis and recommendations. To enhance the accountability function of the CESD, the SCESD has a session each year on the Report and then calls witnesses from each department that is not performing to explain its lack of progress. This is intended to send a signal to both the department and its minister to make improvements; however without any real consequences for inaction the impact of these hearings has been marginal.[42]

The Commissioner's Perspective is the overview chapter of each Report which articulates the CESD's view on Canada's engagement on SD. It provides a concise description of the key messages of the Report's chapters and of the state of SD in Canada. Over the years the Perspective has become increasingly critical of the federal government's efforts towards SD, in particular the lack of engagement from the executive and continual disappointment with the SDS process. In the 2006 Report the Commissioner writes, "I am more troubled than ever by the federal government's long-standing failure to confront one of the greatest challenges of our time. Our future is at stake."[43] While these reports are well-received and considered by parliamentarians and the media and draw attention to important issues, at least in the immediate days following their release, there are currently no mechanisms in place to ensure the recommendations are implemented.

MAJOR ERAS

Inception – Brian Emmett, June 1996 to January 2000

Brian Emmett worked for EC for many years and was a key player in the development of the Green Plan. He left EC in 1994 to become managing director of the Propane Gas Association of Canada in Calgary, but returned to Ottawa in 1996 to take on the position of Commissioner. Establishing the office and position of the CESD was a daunting task, especially given the challenge of reconciling the office's advocate/auditor functions. In the early days, Emmett treated the office as a "mini-university" to generate thinking on SD within government. The primary tools used in this era were studies and polling departments, with an added focus on building relationships and establishing the CESD's profile. In effect, Emmett was pioneering to put departments on a sustainable track.

Over time it was discovered that the studies, though useful to departments, were not garnering attention from the public or media and thus were not helping increase recognition of SD or the CESD. As the CESD was struggling to establish itself and its relevance to SD, public and media attention was highly valued. The success of the 1999 toxic substances audit, which attracted parliamentary hearings and media attention, led to a shift away from studies and greater engagement by Emmett with the audit process.[44]

Although a great deal of momentum seemed to exist in the early stages of the SDS process, it soon faded to frustration and apathy. After a short time, Emmett found that departments were not delivering on commitments and were "playing games"; there was simply no interest in implementing. Efforts to establish relationships with parliamentarians were bearing little fruit. This lack of engagement with parliamentarians would become an important focus for the next commissioner. In 2000, Emmett left for a position at the Canadian International Development Agency. In the end, he was more of a policy person than an auditor, but was instrumental in establishing the credibility of the CESD. Indicative of the dynamic and confusion within the CESD then and now, the right person was hired for a commissioner position but not for an environmental auditor one.

Strengthening and Increasing It's Profile – Johanne Gélinas,
August 2000 to January 2007

When AG Denis Desautel appointed Johanne Gélinas as commissioner in 2000, both the position and methodologies of the CESD were well established. Gélinas' mandate was to implement the Act, build communications and raise the profile of the CESD. Gélinas also wanted to build a better relationship with public servants so that the CESD was not seen as a "bad guy," but as an ally in helping departments to achieve their commitments. Early on she realized the importance of developing a strong relationship with parliamentarians from all parties and made it one of the priorities of her job. She tried to diversify her engagement with committees, to expand beyond the SCESD to Senate and other House committees. While there were formal mechanisms for this engagement such as hearings and meetings, Gélinas also put a great deal of effort into informal relationships as well. Certain House and Senate Committees quickly became her greatest allies and were pivotal to raising the profile of the office and breathing further life to her annual reports. Some of the issues raised in the reports made their way into Question Period in the House of Commons and parliamentarians used the reports as evidence when challenging each other on SD issues. Gélinas also performed numerous speaking engagements to increase the salience of SD with the public and highlight the work of the CESD.

Gélinas had real concerns with the SDS process. She felt that the reporting was done poorly and handled mostly by consultants, indicating that departments were not taking SD into the core of their operations. It was clear that the SDS process was inherently flawed, but despite the urging of Gélinas and numerous recommendations from those within the SD community the oft requested federal strategy to support the process was not forthcoming. The petitions process, on the other hand, was proving to be an important tool for public engagement and awareness-raising, albeit highly underused. In order to strengthen the results and accountability of the petitions process Gélinas initiated a yearly audit of departmental responses to petitions. She also used the annual reports to the House to their utmost potential, putting considerable effort into ensuring they were written in plain language and understandable to the general public. In particular, the Commissioner's Perspective provided a (somewhat) uncensored account of SD in Canada; expressing Gélinas' frustration with the lack of commitment and action from both the departments and the central agencies. Her relationship with parliamentarians and keen sense of timing resulted in the great success of the climate change report in 2006. Although the audit of climate change was initially intended to be only one chapter of the report, Gélinas felt its salience with parliamentarians at the time required that it be the subject of the full report.[45]

An interesting and yet less well known success story during the Gélinas era was the CESD's international work on behalf of the OAG. In 2001, the CESD entered the international sphere with the International Organization of Supreme Audit Institutions (INTOSAI), a UN body with 184 member countries. The AG was appointed chair of the environmental working group for INTOSAI, a position that was delegated to the Commissioner. Thanks to the dedicated efforts of Gélinas and the CESD team, the environmental working group is widely considered the most successful committee of INTOSAI. The group has accomplished much for environmental audit worldwide and has firmly established Canada as a leader in environmental audit. Initially having only 15 member countries in the working group, by the end of Gélinas' six year term as chair there were 60. Countries were attracted to the group because it afforded the opportunity for countries to work in an open and collaborative way, without pressure.[46] Domestically, however, attempts to coordinate with the provincial governments have not been nearly as successful. Federal and provincial audit groups gathered together four years ago for two half-day meetings on environmental audits. There is a third meeting on the agenda for sometime in the future, however there is currently no driving force for such collaboration. The CESD has a closer relationship with certain provinces, such as British Columbia and New Brunswick, but overall there has been a general lack of attention to collaboration and coordination on SD domestically.

Within a few months of Gélinas taking on the position of Commissioner Sheila Fraser became the new AG, bringing with her both organizational and product changes. Fraser put an emphasis on quality management and made plans for greater integration of the SD and audit functions within the CESD.

According to CESD Principal John Reed, this resulted in greater innovation and more teamwork within the CESD. However, despite the progress made on integrating the CESD within the OAG, disagreement between Gélinas and Fraser on the role and limitations of the commissioner increased over the years and came to a head during the last two years of Gélinas' tenure. The AG established a committee that required each group within the OAG to present their audit project plans for analysis. A committee of peers within the OAG would then verify methodology, rigour and general quality control. However, the committee soon began to try to influence topics chosen for the commissioner's report and outcomes.[47] This intrusion into the autonomy of the commissioner only increased the strain on the professional relationship between Gélinas and Fraser. While Gélinas preferred that the CESD have more of an advocacy role in the future, Fraser fundamentally disagreed. The differences in philosophy between the AG and the commissioner resulted in Gélinas' abrupt dismissal in January 2007.

The era of Johanne Gélinas increased the profile of the CESD and helped to maintain environment and SD on the radar of the government and the media. In an interview with CBC, NDP environment critic Nathan Cullen stated "Madame Gélinas has provided the environment community and Canadians with some of the finest work we've seen … she has held the government to account."[48] Each commissioner has the ability to choose how they want to proceed and Gélinas chose to push the limit within the legal mandate of her position. It is apparent that once again the individual chosen to lead the CESD was ideal for a commissioner function and less so as an environmental auditor. The dismissal of Gélinas, along with the universal disappointment with the SDS process, opened the floodgates in the media and Parliament, calling into question the current role of the CESD and the constraints of working within the OAG.

Reflection – Ron Thompson (interim Commissioner),
February 2007 – May 2008

Ron Thompson had been an assistant auditor general for 22 years before Sheila Fraser asked him to fill in as interim commissioner. His tenure was dominated by an inward look at the CESD in response to criticism from the media and parliamentarians following Gelinas' departure. In July 2007 the AG created an independent "Green Ribbon Panel" (GRP) consisting of Elizabeth Dowdeswell, (chair), Jim Mitchell, and Ken Ogilvie, who were tasked with examining "how the mandate established by Parliament in 1995 has been put into practice and to identify any opportunities to strengthen its implementation to better serve Parliament."[49] The GRP report argued that over the past 12 years the CESD has had a positive impact on the federal government's management of environment and SD issues and had developed a strong reputation as a center of excellence in environmental auditing. Although the report did highlight both sides of the debate – those who favour the current model (located within the OAG) and those who prefer and independent CESD – its recommendations were limited to identifying how to strengthen the position as it currently exists. The Panel concluded that there is "considerable room for a more complete expression of the Commissioner's existing mandate, in ways that would better serve Parliament and that would address many of the concerns expressed by the people with whom we spoke."[50] However, it is unlikely that recommendations that include doing more of the same, bringing back studies, and limiting the CESD's term to seven years will go far enough to address the serious challenges that this institution faces in achieving greater implementation of SD.

Thompson dedicated the October 2007 CESD report to examining the effectiveness of the CESD's two main tools – the SDS and petitions process. This report detailed numerous problems with the SDS process and once again called for a federal SD strategy along with a government-wide review of the SDS process. The petitions process on the other hand was characterized as a "good news" story and recommendations were made to make Canadians more aware of the process and to provide better guidance to petitioners. Both the 2007 commissioner's report and the GRP report have been framed with the goal of improving on what currently exists, however there is growing interest in Parliament regarding the role of the CESD; what it is, what it could be, and how it should fit into Canada's broader SD governance system.

In June 2008, Ron Thompson retired from the OAG and Scott Vaughn was appointed the third CESD. The new commissioner arrived at the same time as the new *Federal Sustainable Development Act* (FSDA) gained Royal Assent, ushering in a new era for the CESD – one that may perhaps bring about positive change and renewed vitality to this institution.

BARRIERS AND CHALLENGES

Implementation Gap

The persistent gap between rhetoric and implementation continues to be a large barrier to putting Canada on a sustainable path. Weak implementation limits the audit function by making it difficult to tell what works and what does not. The implementation gap has also blunted the impact of many viable initiatives departments have proposed. When departmental officials include innovative ideas in their SDS but rarely see their ideas translated into action, the disappointment makes innovation less likely in the next round of strategies. The paucity of senior management engagement and incorporation of SD goals into departmental business plans has been a major barrier and perpetuated this problem.

Location within the OAG

Perhaps the most prominent challenge of the current structure of the CESD is its location within the OAG. As Charles Caccia stated, "SD is too important to relegate to the OAG."[51] The commissioner's rank as a second-level official provides too low a profile for such an important issue. Although the OAG is a familiar institution with a solid reputation, the CESD's lack of autonomy from the AG has greatly restricted both the type of work the CESD can do and its internal capacity building.[52] The role of an auditor necessitates a dispassionate and non-policy orientation. As such, the CESD necessarily cannot actively advocate innovative policies that would help put Canada on a sustainable path. As a result, SD policies are left to the guidance of individual ministers; deferring the long-term guidance on SD to those with short-term political life spans. The only tool with the potential to advocate best practices are the studies; however the OAG's discomfort with their policy-oriented nature in the past has made the audit office an inhospitable home. The absence of a forward looking advocacy function and the inability to issue early warnings or suggest issues of importance has resulted in a stagnant process. Cohabitation with the OAG has presented internal challenges as well, as there still exists tension around this "new" component within the OAG. Even within the CESD itself the SD and audit functions can sometimes be at odds, "The CESD represents two different worlds ... sometimes actors within have the impression that they are competing with each other."[53]

Restricted Toolbox

Although the audit function of the CESD is second to none, audit has proven to be a necessary yet insufficient aspect of environmental and SD governance. Audits provide a rear view perspective and little in the way of direction to a department; "performance audits are not intended to make value judgements regarding the relative merits of government policies. Rather, they examine the government's management practices, controls, and reporting systems, based on the government's own public administration policies and on best practices."[54] Being reactive, they are not a tool that can be used to guide future progress. As a "stick" they have the potential to encourage departments to improve performance with some initiatives; however, they are currently ineffective due to the lack of consequences if departments do not meet their commitments. A significant gap in the CESD's audit process is that it performs audits on whether commitments are implemented and not on whether the strategies proposed are effective or appropriate to achieving SD. One would think that this would be the most important consideration – as implementing an ill-conceived strategy does little to help Canada's sustainability efforts.

The other tools in the CESD's arsenal have been found to be relatively weak. As noted earlier, the SDS process is considered by most to be a paper-pushing exercise that is not taken seriously by departments or parliamentarians. The lack of incentives and disincentives and low-level engagement by decision-makers present barriers to the success of the strategies. The reliance on the low-impact "name and shame" approach has proven ineffective at compelling improvement. As well, some departments that are essential to making Canada sustainable, for example the Department of Finance, are not accountable to the CESD making the overall effort ineffective. While the petitions process has been successful in providing a mechanism for public participation and awareness-raising, it is so little used that its influence is limited. The commissioners reports have been excellent in raising awareness of important issues and increasing accountability; however, the effects of the spotlight are only temporary and have not promoted change from business as usual.

Lack of Executive Leadership

Without a strong overall commitment from the executive branch, there is little motivation for those with the power to implement SD programs and policies to do so. While a number of departments and institutions may be responsible for different aspects of SD, none is responsible for the whole. Without a federal strategy, the rear view of the audit process is not being complimented by a forward-thinking and cohesive policy approach. Currently it is up to individual parliamentarians to take audit findings and examine if their policies are wrong, which assumes that each parliamentarian will have the expertise and understanding to determine the most appropriate and sustainable policies. This is a seemingly haphazard and stove-piped approach to what has been characterized as one of the greatest challenges facing humanity. To help counter this, the new FSDA requires that the government create a federal strategy, which has been tasked, not surprisingly, to EC. It is unclear the direction that EC will take with this strategy, or whether or not an environment minister will be able to effectively incorporate economic and social considerations into the strategy. As well, the language of the Act focuses on "environmental integration," which reinforces the fact that Canadian leadership is not taking to heart the three pillar approach that is the essence of SD.

There is also a general lack of SD leadership within the senior management levels of the bureaucracy. According to the 2008 CESD report, instances where federal departments were successful in effecting change through their SDS, senior managers were involved in the articulation of policy; they communicated a clear message, provided technical support and guidance, marshalled necessary information, and established performance monitoring systems. Where federal SD governance initiatives have fallen short has been because elements of this management approach were missing.[55]

Institutionalizing a SD approach within Canadian governments confronts the challenge inherent in all horizontal issues that apply across sectors, as government departments are organized by sector and are mandated to protect and promote the interests of their individual sectors, which may not be aligned with SD objectives. This "horizontal management gap" is pervasive in most federations and democracies. The executive in Canada has not yet thought through how to implement an SD approach horizontally, but it must if SD is to infiltrate the governance system. Central agencies should have a strong role in the coordination of SD efforts across sectors; however, these agencies (PCO, TBS, Finance) do not currently have the expertise nor the explicit mandate to do so. As a result, central agencies consider SD a "departmental thing"; while many departments feel that it is not their mandate either.[56] The CESD has tried several different approaches and messages to get the central agencies and executive more engaged, but to no avail.

Advocate/Auditor Dichotomy

Many of the challenges discussed thus far arise from the fact that the title of commissioner arouses certain expectations of what that role is supposed to deliver. The commissioner title was chosen because the House of Commons "wanted environment to be special" but was accompanied by the mandate of an auditor.[57] This simple misnomer has resulted in a great deal of confusion; in Parliament, the public and within the OAG itself. As AG Sheila Fraser has stated: "Based on the bills which have been brought to our attention ... there seems to be a discrepancy between expectations being placed upon us and what we are actually in a position to do, and people are not aware of this."[58] While a commissioner typically acts as an advocate and advisor and comments on the policy direction of the government (such as with the commissioner on Human Rights), an auditor typically fills in the accountability and information function. The fact that many are pushing for the CESD to take on more of an advocacy role, including the latest commissioner herself, demonstrates a significant gap that exists in the system.

ACHIEVEMENTS AND OPPORTUNITIES

Focus on Accountability

The commissioner performs an important role in the SD system in Canada, as it is the only body whose sole purpose is to look at implementation. In its mandate to hold government accountable, the CESD has been successful. The CESD's audit capacity is internationally respected and has been pointed to as a model by the OECD. As well, the CESD reports have provided an unbiased assessment of SD in Canada. Although more difficult to measure, the CESD has had an influence on departments as well, with changes made to decision-making process, increased legislation, and increased funding to work on specific sd issues.[59] There are also linkages between funding in the federal budgets and issues that were raised in cesd reports. Indeed, several key clauses of the fsda are direct outcomes of "lessons learned" from over a decade of cesd reports and experiences.

Engagement of the Public

The cesd has also been successful in helping maintain environment and sd on the radar screen of the government, media and public. The petitions process is providing a much-needed avenue for public participation. The potential reincarnation of the study as a tool of the Commissioner could provide another opportunity for improvement and increasing innovation within the cesd. Although there will be barriers to overcome within the oag regarding the content and style of the studies, they are poised to serve an important function in promoting best practices and knowledge generation both inside and outside of government.

International Audit Assistance

The impact of the cesd internationally has been considerable. As chair of the intosai environmental working group, the commissioner has supported the efforts of the oag-equivalent worldwide to perform environmental audits. This working group is furthering the development of cohesive environmental audit practices internationally. When countries around the world work together and take a similar approach to environmental auditing it can really help advance policy learning on an issue. It will be easier to support action and assess which policies are working if five other countries perform environmental audits on the same issue. Many of the members of the working group are also developing countries, where sd is particularly important in helping them to avoid the destructive development path that has lead to many of the world's current environmental and social problems.

Parliamentary Interest in Sustainable Development

The effort Gélinas put in to forming relationships with parliamentarians has also born fruit. It has helped to engender in some parliamentarians a real passion for the issue and momentum to address some of the challenges the cesd currently faces. Since the high profile removal of Gélinas, there has been a great deal of debate over the institutional structure of the cesd and its autonomy from the oag. Recent motions and bills have been put before the scesd and the House of Commons to change the current legislation. In

2007, Liberal MP David McGuinty put forward a motion to the SCESD to examine how to strengthen the role of the CESD. After a thorough review the SCESD called for the government to enact legislation making the Commissioner an arms length, full and independent agent of Parliament.[60] These recommendations were presented to the House on March 1, 2007, however they did not result in further action. More recently, Liberal MP John Godfrey proposed Private Members Bill C-474:

An Act to require the development and implementation of a *National Sustainable Development Strategy*, the reporting of progress against a standard set of environmental indicators and the appointment of an *independent Commissioner of the Environment and Sustainable Development accountable to Parliament*, and to adopt specific goals with respect to sustainable development in Canada, and to make consequential amendments to another Act.[61]

The Bill received Royal Assent in June 2008; however the reference to the creation of an independent commissioner was withdrawn after being ruled to be out of order by the Speaker of the House. The creation of an independent commissioner would have required a Royal Recommendation by the government and the current Conservative government was not prepared to create an independent office. Still, the Act is a promising development and demonstrates the continued attention to the importance of SD as an issue, although the full implications of the Act are still unknown.

These endeavors demonstrate the continued salience of SD and indicate that perhaps change is in the air. The new Commissioner Scott Vaughan will be able to set his own direction for the CESD, within the constraints of the current mandate. Commissioner Vaughan will be taking over a mature and well run audit office and will have the benefit of the recent evaluations, reports and discussions in Parliament. As with any transition, there will be an opportunity for change and rebuilding and to strengthen what currently exists.

MOVING FORWARD

Despite its poor record, Canada still has the potential to be an international leader on SD. Canada has a natural comparative advantage given its diverse landscape and climate, vast experience working with natural resources, world renowned educational institutions, and leadership in the development of innovative technologies. The Canadian public is also ready. A 2007 poll showed the majority of Canadians believe it is the responsibility of the government to "fix environmental problems" and even more felt that individuals need to take an active role as well.[62]

Unfortunately, to date Canada's institutional engagement with SD has been a failure and the country continues to develop in the direction of consumerism and unsustainable economic growth. Although 12 years is not a long time for an institution to become established, nor 21 years for bringing about transformative change throughout society, progress towards SD in Canada has stagnated. Many of the troubling premonitions of the 1994 SCESD recommendation report about the outcomes of placing the commissioner within the OAG have come to pass. Most of the challenges facing the Canadian system have arisen from lack of advocacy, policy support, and independence, all things which a commissioner should be able to provide. A movement for change towards a sustainable future should include the following ingredients:

Sustainable Development Auditor

It is important when implementing change to fix only what needs fixing and to strengthen what is already working well. The CESD has developed a world leading environmental audit capacity, which is well supported by the OAG. The creation of an SD Auditor (SDA) position would enhance the current audit function by increasing the audit focus and removing the policy-type functions that have caused confusion both within the OAG and externally. The SDA would retain much of the current mandate of the Commissioner, including performance audit, auditing of special issues and petitions. It would also build the capacity to audit for SD itself, making it possible to measure the effectiveness of departmental initiatives on Canadian society so that they may improve over time. The SDA would work in collaboration with an independent Commissioner and would produce an annual joint report to Parliament. While the SDA would continue to highlight special issues and hold departments accountable, the commissioner's role would be to elaborate best practices and suggest future directions.

Independent Commissioner of the Environment and Sustainable Development

An independent CESD would be able to propose direction to Parliament and have an important policy advisory role. He or she could: independently determine what issues to explore; research best practices; anticipate problems; and undertake special studies to show how SD has been addressed in the private sector or other levels of government. Not being restricted by the audit culture of the OAG, the CESD could use a wider variety of tools and explore new innovative ideas. Former Commissioner Gélinas stated in recent testimony to the SCESD:

A commissioner must be able to offer a vision, an approach, a way of acting and a general orientation. He or she must be able to debate, to promote activities, to work with departments in other ways than simply through audits.

Although widely used and even essential, the audit tool cannot all by itself create change within the government administration. In the area of sustainable development, what is needed above all is education and collaboration, not solely auditing. And the very nature of the Office of the Auditor General of Canada does not permit this kind of work …

The model that best corresponds to the creation of an independent office for an environment and sustainable development commissioner is that of the Commissioner of Official Languages. This position is independent and reports directly to Parliament. Although the Official Languages Commissioner does do audits, there are other tools at his or her disposal to advance the cause of bilingualism.[63]

Through studies, advocacy, and innovative policy support, an independent CESD could help fill the gaps that currently exist and bring balance to the rear-view approach to SD that has been established in Canada. The commissioner would also be aware of, and perhaps collaborate with, initiatives on-going in Canada's other SD institutions (NRTEE, IISD, SDTC), the provinces, and civil society.

Executive Engagement and Prioritization

The federal government plays a pivotal role in setting the agenda on SD issues and establishing SD as the foundation of Canadian society, "As Canada's largest single employer, landlord and purchaser, the federal government can show leadership in taking due action."[64] According to former EC Senior Assistant Deputy Minister Robert Slater, the best way to make the federal government sustainable is through leadership, learning and accountability. The recent inquests into the CESD and its tools have increased our learning – and we have learned that we are not delivering on the leadership and accountability aspects of SD.[65] The prime minister and cabinet have an essential role to play in increasing the implementation of SD initiatives through creating internal incentives to promote institutional change and ensure monetary and career repercussions for failing to meet SD-related commitments. If SD considerations were included in the performance pay of deputy ministers, there would most likely be much more progress on SD than we see today. Increasing consequences would also strengthen the influence of the SDA and the CESD, as parliamentarians would be more likely to follow up on their recommendations. Indeed, the FSDA includes such a provision on senior executive performance contracts.

Before any progress can be made, the executive branch needs to seriously think through how it wants to implement SD; including how to horizontally infuse a SD approach to policy and programming through all departments and, especially, central agencies. Central agencies will be essential in coordinating action across departments, which would best be facilitated by the addition of a group focussed specifically on sd within the pmo and Privy Council Office. The executive also needs to get rid of perverse policies that impede sd by integrating the environmental and economic agendas, "the Government must show leadership and accountability in integrating environmental and economic decisions."[66] Part of the process of integration will be ensuring that departments incorporate sd commitments in their budget and planning documents and provide extra resources for innovative initiatives. Political leadership across all parties is essential for Canada to make progress on its goals. As Brundtland stated in the opening statement to ocf, "in the final analysis sustainable development must rest on political will."[67] A federal strategy, that is reviewed and supported by a cesd and audited by the sda, would provide government departments with a common vision and definition of sd and common goals to base their sdss on: "The government has indicated that sustainable development is a government-wide initiative, not just a departmental one. Therefore, in order to bring departmental strategies to life, the government needs to clearly articulate what its sustainable development goals are, for the government as a whole, and how individual departments are expected to contribute to achieving them."[68] By reviewing subsidies and implementing ecological fiscal reform, the strategy could help the government to overcome some of the greatest barriers to sustainability in Canada.

CONCLUSION

The office of Commissioner of Environment and Sustainable Development has achieved some real success in its current mandate. However, the experience of the past 12 years has demonstrated that audit is not enough to put the Canadian government on a sustainable track. All three cesd's have reiterated time again that the federal government and its departments are not living up to their promises; however this awareness has done little to combat the source of the problem. The accountability provided by audit is an important but not sufficient component of sustainable development governance. Missing is an overall direction and a coordinated approach to sd in Canada, one that takes into account *all three* pillars of sd. An independent cesd could provide Canada with the forward-thinking advice and advocacy necessary for sd integration, complimented by the information and assessment of an sd auditor. The bottom-up approach of departmental strategies would be reinforced by the guidance and a vision that a federal sd strategy could provide. Each of these components would necessarily be supported by leadership from the executive and central agencies, as they will determine if sd is a priority in Canada. Canada needs to set up a structure that will outlive changes in politics; one that innovates, coordinates, and compels.

NOTES

1 William Lafferty and James Meadowcroft, "Patterns of Governmental Engagement," in William Lafferty and James Meadowcroft, eds. *Implementing Sustainable Development: Strategies and Initiatives in High Consumption Societies* (Oxford University Press, 2000), 435.

2 Canadians Jim McNeill and Maurice Strong were involved in the creation of OCF as members of the World Commission on Environment and Development, which held numerous consultations across Canada in 1986.

3 Ken Ogilvie, personal interview, 29 November 2007.

4 Glen Toner, "The Green Plan: From Great Expectations to Eco-backtracking … to Revitalization?" in Susan Phillips, ed. *How Ottawa Spends 1994–95: Making Change* (Carleton University Press, 1995); Ogilvie, 2007.

5 Wayne Cluskey, personal interview, 14 August 2007.

6 Stratos Inc., *Governance Models for Sustainable Development* (Policy Research Initiative, August 2002), 2.

7 Tim Williams, *Sustainable Development in the Federal Government: The Commissioner of the Environment and Sustainable Development* (Library of Parliament, July 2005), 2.

8 Cluskey, 2007.

9 Robert Slater, personal interview, 14 August 2007; Rick Smith, personal interview, 12 February 2008.

10 Glen Toner, "Environment Canada's Continuing Roller Coaster Ride," in Gene Swimmer, ed. *How Ottawa Spends 1996–97: Life Under the Knife* (Carleton University Press, 1996), 109.

11 The SCESD is where the main parliamentary work on SD is done. It addresses SD generally under specific provisions including the Canadian Environmental Assessment Act (2003 amendments*), the National Round Table on the Environment and Economy Act (1993), the Auditor General Act (1995 amendments), the Oceans Act (1996) and Canadian Environmental Protection Act (2005*), as well as various statutes establishing departments. Frédéric Bouder, *Governance for Sustainable Development in Canada* (OECD, 2001), 15. *updated.*

12 Ron Thompson, personal interview, 26 July 2007.

13 Standing Committee on Environment and Sustainable Development (SCESD), *Report on The Commissioner of the Environment and Sustainable Development.* House of Commons Issues No. 27 (May 1994), xi, 1.

14 Ibid., 3.

15 Ibid., 6.

16 Ibid.

17 Brian Emmett, personal interview, 22 August 2007.

18 See chapter 3 of this volume.

19 Stratos, *Governance Models*, 9.

20 Office of the Auditor General of Canada (OAG), "A Brief History" (2007). Available at: www.oag-bvg.gc.ca/domino/cesd_cedd.nsf/html/cesd_hist_e.html; accessed 25 March 2007; Cluskey, 2007.

21 Elizabeth Dowdeswell, Jim Mitchell, and Ken Ogilvie, *Fulfilling the Potential – A Review of the Environment and Sustainable Development Practice of the Office of the Auditor General of Canada* (Green Ribbon Panel, January 2008), 12. Available at: www.grp-gev.ca/reprap-e.htm.

22 OAG, "Petitions Catalogue" (2007). Available at: www.oag-bvg.gc.ca/domino/petitions.nsf/viewE1.0?openview&count=1000; accessed 18 January 2008.

23 Dowdeswell et al., *Fulfilling the Potential* (2008).

24 Canada, *A Guide to Green Government*, (1995); House of Commons, *Bill C-474: First Reading 5* (2nd Session, 39th Parliament, November 2007). Available at: www2.parl.gc.ca/HousePublications/Publication.aspx?DocId=3096584&Language=e&Mode=1.

25 Katie DeRosa, Harris MacLeod, and Bea Vongdouangchanh, "PM voted most valuable politician in 2007: survey," *The Hill Times*, 17 December 2007.

26 Luc Juillet and Glen Toner, "From Great Leaps to Baby Steps: Environment and Sustainable Development Policy Under the Liberals," in Gene Swimmer, ed. *How Ottawa Spends 1997–1998: Seeing Red- A Liberal Report Card* (Carleton University Press, 1997), 195.

27 Charles Caccia, personal interview, 5 February 2008.

28 Standing Senate Committee on Energy, Environment and Natural Resources, *Sustainable Development: It's Time to Walk the Talk* (June 2005). Available at: www.parl.gc.ca/38/1/parlbus/commbus/senate/com-e/enrg-e/rep-e/repintjun05–e.htm.

29 Johanne Gélinas, personal interview, 21 February 2008.

30 John Reed, personal interview, 2 August 2007.

31 Williams, *Sustainable Development in the Federal Government*, 3.

32 Reed, 2007.

33 Commissioner of the Environment and Sustainable Development (CESD), *Report of the Commissioner of Environment and Sustainable Development to the House of Commons* (October 2007), 22.

34 Ibid., 7.

35 Gélinas, 2008; Emmett, 2007.

36 Slater, 2007.

37 Auditor General, *October Report of the Office of the Auditor General: Matters of Special Importance – 2007*. Available at: www.oagbvg.gc.ca/internet/English/parl_oag_200710_00_e_23824.html; accessed 15 November 2007.

38 Industry Canada, Natural Resources Canada, and Transport Canada are among the few that have actually integrated their SDS planning into internal management processes. Still, many of the proposed "SD" plans are modest or initiatives that they would have done anyway. See François Bregha, "Missing the Opportunity: A Decade of Sustainable Development Strategies," in Glen Toner, ed. *Innovation, Science, Environment: Canadian Policies and Performance 2008–2009* (McGill-Queen's University Press, 2008), 30–52.

39 Example of study topics were *Expanding Horizons – A Strategic Approach to SD* (1998), *Working Globally – Canada's International Environmental Commitments* (1998), *Building a Sustainable Organization – The View from the Top* (1999), *Working with the Private Sector* (2000), and *Integrating the Social Dimension – A Critical Milestone* (2001).

40 Thompson, 2007; Smith, 2008; Gélinas, 2008.

41 Glen Toner and Carey Frey, "Governance for Sustainable Development: Next Stage Institutional and Policy Innovation," in G. Bruce Doern, ed. *How Ottawa Spends 2004–2005: Mandate Change in the Paul Martin Era* (Carleton University Press, 2004), 215.

42 Caccia, 2008.

43 CESD, "The Commissioners Perspective," *Report of the Commissioner of the Environment and Sustainable Development to the House of Commons* (2006). Available at: www.oag-bvg.gc.ca/internet/English/parl_cesd_200609_00_e_14982.html.

44 Neil Maxwell, personal interview, 12 February 2008.

45 Thompson, 2007.

46 Thompson, 2007; Gélinas, 2008.

47 Gélinas, 2008.

48 "Canada's Environment Commissioner: Firing a "complete surprise,"" CBC *News*, January 30, 2007. Available at: www.cbc.ca/canada/story/2007/01/30/gelinas-fired.html.

49 Dowdeswell et al., *Fulfilling the Potential*, 43.

50 Ibid., 10.

51 Caccia, 2008.

52 Glen Toner, Testimony before the SCESD, 12 February 2007. Available at: www2.parl.gc.ca/HousePublications/Publication.aspx?DocId=2687836&Language=E&Mode=1&Parl=39&Ses=1#Int-1893109.

53 Gélinas, 2008.

54 Williams, 3.

55 CESD, *Report of the Commissioner of Environment and Sustainable Development to the House of Commons* (2008), 3.

56 Stratos, *Governance Models*, 14.

57 Slater, 2007; Thompson, 2007.

58 Sheila Fraser, Testimony before the SCESD (2007). Available at: http://cmte.parl.gc.ca/Content/HOC/committee/391/envi/evidence/ev2679902/enviev41–e.htm#Int-1887907.

59 Smith, 2008.

60 SCESD, *Second Report to the House of Commons* (39th Parliament, 1st Session, 2007). Available at: http://cmte.parl.gc.ca/cmte/CommitteePublication.aspx?SourceId=191556.

61 House of Commons, *Bill C-474: First Reading*, 2007.

62 Ipsos Reid, "Canadians and Their Governments Both Responsible for Fixing Environmental Problems," 4 July 2007. Available at: www.ipsos-na.com/news/pressrelease.cfm?id=3557; accessed 8 May 2008.

63 Johanne Gélinas, Written testimony to the SCESD, February 26, 2007.

64 Frédéric Bouder, *Governance for Sustainable Development*, 13.

65 Slater, 2007.

66 SCESD, *Second Report*, 1.

67 World Commission on Environment and Development, *Our Common Future* (Oxford University Press, 1987), 9. Available at: www.worldinbalance.net/agreements/1987–brundtland.html.

68 CESD, *Report to the House of Commons* (2007), 8.

12 Building a Sustainable Development Infrastructure in Canada: The Genesis and Rise of Sustainable Development Technology Canada

ANIQUE MONTAMBAULT

Sustainable development (SD), as popularised by the World Commission on Environment and Development (also known as the Brundtland Commission) definition in *Our Common Future*, entails meeting the needs of the present without compromising the ability of future generations to meet their own needs. This will involve a radical departure from our current modes of production and consumption, and provides an enormous opportunity for Canada, rich in natural resources, to find new processes to protect, conserve and utilise our natural capital. There are several Canadian companies that are on the cutting edge of research and development of environmental technologies. Yet, they face numerous challenges in brining these technologies to market, such as development of the product, financing, or successfully marketing their product. The assistance, guidance, and financial support from Sustainable Development Technology Canada (SDTC), is instrumental in these technologies' path to market.

Created in 2001, SDTC is a foundation whose purpose is to "provide financial support to projects that develop and demonstrate new technologies that have the potential to advance SD, including technologies to address climate change, clean air, and water and soil quality issues."[1] Its $342 million investment in technologies has generated $800 million in leveraged funding. Prior to SDTC, early stage, or seed, equity investment in these environmental technologies amounted to less than four percent of the total investment in Canada due to the high technological, financial and market risk faced by these technologies. SDTC has successfully bridged the funding gap for environmental technologies in Canada. This chapter will trace the origin and evolution of this foundation, examine why this model was chosen, evaluate its challenges and opportunities, and address the possible future for SDTC.

GENESIS: DRIVERS IN THE CREATION
OF SDTC

Kyoto and the National Climate Change Process

Canada was an international pioneer of SD, as the stories of the development of the International Institute for Sustainable Development (IISD) and National Roundtable for the Environment and the Economy (NRTEE) described in the previous chapters show. The 1986 visit to Canada of the Brundtland Commission, the concern over ozone depleting substances and the atmosphere leading to the 1988 Toronto conference, and a strong performance at the Rio Earth Summit firmly established this reputation. Both the Conservatives and the Liberals affirmed their commitment to SD principles by placing it at the core of key policy documents: the Progressive Conservative's 1990 Green Plan, the Liberal's 1993 electoral manifesto known as The Red Book and their subsequent directive to government departments, the Guide to Green Government. This strong political involvement in SD and climate change discussions created the momentum leading into the 1997 Kyoto negotiations, where Canada committed to a 6 percent reduction of greenhouse gas emissions on 1990 levels for the 2008–12 period. While the political will was there to commit to an international agreement, the solutions to climate change were not as evident; as the Commissioner of Environment and Sustainable Development's 2006 Report noted, the decision to match or exceed the reductions of its main trading partner, the United States, was made in the absence of any economic analysis or social and environmental risk assessment.[2]

As a consequence of the federal and provincial conflict over how to achieve the Kyoto commitment, federal and provincial energy and environment ministers created the National Climate Change Process (National Process) to assemble the best knowledge and policy options for the Liberal's new policy on climate change. This was an in-depth, two-year long, national consultative project from 1998 to 2000 involving 450 experts from government, industry, academia, non-governmental organisations. One of the sixteen "issue tables," the Technology Issues Table, had the dual mandate to "propose ways to advance the development and commercialisation of cleaner and/or innovative technologies which will contribute to the reduction of greenhouse gas emissions; and enhance the capabilities and opportunities for Canadian companies in providing environmentally-responsive technologies in domestic and international markets."[3] In their report, the co-chairs noted that a long term focus was necessary considering the timeframes in which technologies are developed, from conception to market, and that this required an analysis of the "means to develop the intellectual and institutional capital to ensure technologies and systems are available"[4] (emphasis added). Of their eight recommendations, one proposed the creation of an $840 million Climate Change Technology Demonstration Program. Its purpose would be to support technological demonstrations up to the first full-scale implementation, thereby reducing the risk, and increasing its likelihood of making it to market. Federal government investment would be re-payable and limited to 30 percent of the project – the remaining 70 percent of the funding would be leveraged from industry and other levels of government by encouraging technology user and supplier alliances. For the Program's implementation, the report indicated that the "Technology Issues Table is of the view that delivery of such investments should be largely through existing programs and mechanisms."[5]

Indeed, the need to support the development and demonstration of environmental technologies was recognised within the bureaucracy at least a year earlier when the Technology Early Action Measures (TEAM) program was established within Natural Resources Canada (NRCan) with the objective of funding late stage development and first stage demonstration of technologies with the potential to reduce GHGs. It operated interdepartmentally, with Industry Canada, NRCan and Environment Canada all assuming a role in its strategic direction, and was given a budget of $56 million. TEAM was seen as a trailblazer in government circles, having disbursed funding to companies on average 4 to 6 months after the initial application, and securing a leveraging ratio of 5:1 from other levels of government or industry.[6] It also received the 2000 Head of the Public Service Award for excellence in the policy category. If the bureaucracy had indicated that it was capable, and was even recognised as a leader for the program it created, why did the government create a separate institution?

Linking the Environment and Economy

There were strong drivers to create a separate institution in support of SD technologies. Liberal Cabinet leaders saw an opportunity to integrate the environment and the economy by re-positioning Canada's substantial natural resources sector toward the knowledge-based economy; building capacity in GHG mitigation technology was a natural fit for an energy intensive economy built on natural resources. The announcement of SDTC's creation in Budget 2000 represented a fundamental shift in the government's perception of the environment as an economic opportunity rather than a burden. David McGuinty, then executive Director of NRTEE, commented that the $500 million being committed for a suite of environmental science and technology (S&T) initiatives was the operationalisation of this shift.[7]

Public administration paradigms had also shifted. New Public Management principles of public-private partnerships, cost-recovery, and alternative service delivery became the touchstone for the widespread changes to government operations during the Chrétien government's Program Review.[8] Foundations were a popular mechanism for achieving policy goals. From 1997 to 2000, the government created five foundations, ranging from large with substantial endowments such as the Canada Foundation for Innovation

(CFI) with $3.65 billion to more modest organisations such as the Pacific Salmon Endowment Fund Society with a budget of $30 million. In addition, the negative press attention brought to the government from the 2000 Human Resources and Development Canada scandal also made it less politically appealing to augment departments' core funding. This significant trend toward "autonomisation" is surprising given Canada's reluctance to deviate from strictures of Ministerial responsibility and accountability. The foundation represents a more radical organisational form, and places a significant amount of trust in governors who are beyond the reach of political influence.[9]

NRCan Minister Ralph Goodale argued before the House of Commons Standing Committee on Aboriginal Affairs, Northern Development and Natural Resources (AANR) for the necessity of the foundation model. He described a need for government to be able to approach industry on its own terms, to be "one in the club": "What we're hoping to achieve here is synergy among private sector players and other players outside government ... I think those players deserve a lot of praise for what they've done so far, but we do need a range of tools to further engage them, to continue moving the yardsticks forward. Our view was that having this funding instrument at arm's length from government is a good way to build that synergy and further engage the private sector."[10] He also defended the unique role the foundation could play within the existing innovation system and that its activities were intended to complement, not supplant, existing measures within government:

There are a number of public and private federal and provincial funding processes that deal with the primary research end of the equation, and there are a number of funding initiatives that deal with the commercialisation end of the innovation chain, but there has tended to be a gap in the middle. That's where this fund is aimed, and we've been very careful to make sure we're not overlapping with other initiatives or duplicating what someone else is doing.[11]

This perspective was popular internationally as well. In discussing the instruments for GHG emissions reductions, the OECD had by the mid-nineties argued that governments had a critical and active role to play in stimulating the entire life cycle of environmental technologies, from the development stage to diffusion in the market, as well as a critical role in supporting the basic research which underlies innovation.[12] Therefore, the foundation model was needed as it was better able than traditional government departments to create partnerships between industry and government, and academia.

A Difficult Birth: Creating the Legislation

Two important cabinet members played a strong role in SDTC's creation: Paul Martin, then finance minister, who had attended Rio and authored the

Liberal Red Book's chapter on SD, and Ralph Goodale, chairman of the Economic Committee of Cabinet and lead for the government's implementation of international climate change commitments. Goodale and Martin approached James Stanford, then president and CEO of Petro-Canada, to see how the government could re-orient a portion of the $1.3 billion windfall from the sale in 1995 of Petro-Canada shares toward a specific policy for the environment.[13] At the same time, Goodale convened 15 energy and technology innovation experts outside government, including Vicky Sharpe, the current president and CEO of SDTC, to seek their input on the mandate for an organisation that could address Canada's Kyoto commitments.[14] These experts responded that one of the most significant barriers for Canada to achieve its environmental commitments was the lack of market-ready environmental technologies, and attributed this in large part to the absence of investment from the private sector in the early stages of these technologies' development.[15]

In the early spring of 2000, Prime Minister Chrétien sent a letter to the Ministers' Office of NRCan, Environment, and Industry, directing them to work together to develop the legislative framework to create a foundation for the fall of 2000. NRCan was the lead on this file, and Mary Preville, then manager of the Climate Change Technology Programme, was tasked with developing the drafting instructions for the legislation with the assistance of an interdepartmental working group. The foundation model was based on an existing and successful foundation with similar aims, the CFI. At the same time, NRCan and Environment Canada officials drafted the first funding agreement with the foundations, which spelled out the conditions associated with its funding.

Budget 2000 was the first to publicly announce the creation of the Sustainable Development Technology Fund. Its language drew clear parallels between economic growth, SD and innovation: "the development, dissemination and use of environmental technologies are essential as Canada makes the transition to a more environmentally benign information economy."[16] Bill C-46, *An Act to establish a foundation to fund sustainable development technology*, was introduced on 4 October 2000 but died on the order paper when the general election was called for 27 November.

Following the election, the Department of Finance was however refusing to re-profile the money that had been committed in Budget 2000 if the legislation did not pass by the end of the fiscal year and instructed NRCan to find a way to secure the money by 31 March 2001[17]. To prevent the money from lapsing, and to retain it for the purposes of SD, the government created on 8 March 2001 what Minister Goodale described as an "interim holding company"[18] and the $100 million was transfered to it. Bill C-46 was re-introduced to the House of Commons on 13 March 2001, but modified to include a clause in the legislation to enable the Governor in Council to call upon a not-for-profit corporation under Part 2: corporations without share capital of the *Canada Corporations Act* (CCA), to administer the fund, if necessary, until the Bill was passed. Once the legislation was proclaimed, the Governor in Council could change the corporation into a foundation for the purposes of the Act.

Though the members of the AANR and Standing Senate Committee on Energy, the Environment and Natural Resources (EENR) were widely supportive of the aims of the foundation, there were strong concerns about the transfer of money to a not-for-profit corporation and its accountability to Parliament and the Canadian public. Goodale defended the transfer of money to the "holding company" by saying that legal principles of the government and the CCA had been upheld, and that another foundation had been created in this same way. Furthermore, Goodale emphasised that restrictions were placed on the funding and reviewing of proposals, and the credentials and reputation of the holding company's four founding members – Jim Stanford, Ken Ogilvie, David Johnston, and Alain Caillé – demonstrated the prudence of the government's actions.[19] The Committees' other primary concern was that as a non-governmental agency, the Canada Foundation for Sustainable Development Technology (CFSDT) would be subject neither to the *Access to Information Act*, nor a performance evaluation by the auditor general. Minister Goodale vigorously defended the foundation's accountability structure by outlining the requirement to report annually to Parliament, including detailed project information and a financial breakdown, as well as having to hold a public annual general meeting. However, as David Chatters, MP for the Canadian Alliance party, pointed out to the AANR, the requirement for a private sector auditor, independently chosen by the Board, to conduct an evaluation of the foundation's activities could not provide the same kind of transparency as the auditor general.[20] The auditor general, in response to the EENR, also raised several concerns over the lack of transparency. She noted the inability of the Office of the Auditor General to perform value-for-money audits, and cautioned that her Office's review of the funding agreement between NRCan and the CFSDT could only look at the departmental responsibilities, but had "no authority to look at the operations of the foundation". Further, she questioned the appropriateness of arm's length organisations: "To my knowledge, there has been no study or evaluation done which would indicate that this type of foundation or agency is actually providing anything different from the traditional core government. It would be worthwhile for government to carry out a review of these new, alternative service delivery mechanisms to assess the advantages, disadvantages and what needs to be corrected to move forward."[21]

The opposition to Bill C-4 in the Senate delayed the approval process such that it appeared likely that the bill would not pass before the House rose for the summer. Finally, to shepherd the bill through the Senate, Minister Goodale secured a strong and respected voice in the Senate, Sharon Carstairs, who was then leader of the Government in the Senate. The bill squeaked through at the eleventh hour – 14 June 2001 was the very last sitting day of the House before the summer recess.

Despite the Committees' opposition to the process of the foundation's creation, Goodale pursued C-4 because he argued that specific legislation around SD was necessary to carry out the objectives of the SDTC.[22] This is notable since only two of the sixteen other foundations have been created through legislation: the Canada Foundation for Innovation (CFI) and the Canada Millennium Scholarship Foundation.[23] Providing a legislative basis for the foundation represents a greater step toward the institutionalisation of SD than if the CFSDT had remained a not-for-profit corporation under the *Canada Corporations Act*.

Institution Structure

First created in 1997 with the CFI, the Canadian government has outlined the role of foundations as instruments of public policy: "Not-for-profit organisations governed by independent arm's length board of directors made up of experienced and knowledgeable individuals with expertise in specific areas of research, development and learning. Their arm's length nature, financial stability and focused expertise allow them to address specific challenges in a highly effective non-partisan manner."[24] As outlined in a response to the auditor general's 2005 Report which questioned the accountability of foundations,[25] the government argued the merits of foundations as providing five distinct features that are not possible within, or can be compromised by, traditional department structure. First, they provide focus on complex and challenging issues; otherwise, cross-cutting issues such as SD for which no single department is responsible may flounder as departments disagree over objectives and resource sharing. SDTC was created to fill a niche within a complement of existing government programs and organisations involved in supporting research and development, such as the Industry Canada's Industrial Research Assistance Program (IRAP), the CFI, and TEAM among others. Second, the foundation is built on the experience and expertise of its staff and board members, who work in the applicable field and can provide direction at the board level and input to the peer-review process. Third, the provision of multi-year funding liberates foundations from the annual appropriation cycle and mitigates the risk of on-again off-again funding. In the case of SDTC, this benefits the companies who would not undergo the proposal submission process and risk being denied funding because of a change in political will. Fourth, the guarantee of long-term funding provides the incentive for leveraged contributions from other sectors. And finally, it recognises the long-term nature of scientific research and development which is dependent on stable financial resources.

Like other foundations, SDTC's funding agreements, developed by NRCan and Environment Canada, the sponsoring departments, establish the foundation's requirements to the Government, as well as the parameters for its activities. SDTC's governance framework consists of a board of directors as well as a board of members, which SDTC has called the Member Council. The board of directors governs the business of the foundation and provides the strategic direction. In SDTC's case, the directors are also members of several more operationally focused committees: the Corporate Governance Committee, the Human Resources Committee, the Project Review Committee, and the Audit and Grant Investment Committee. The legislation establishes that the governor in council appoints seven of the fifteen board members, and the remaining eight are selected by the members of SDTC. The Member Council, comprised of 15 individuals, would be drawn from the private, public and academic sectors in Canada to provide stakeholder views to the Board.[26] When the legislation was created, the ratio of government appointments to the Member Council was identical to the Board of Directors; however, eventual vacancies on the Member Council would be filled by selection from the members.

Minister Goodale secured an important group of people for the organisation. Vicky Sharpe was chosen by the government to be the president and CEO of SDTC. She was uniquely prepared to guide SDTC's development: while her formal training at the doctorate level was in microbiology and chemistry as applied to water pollution control, she also had 25 years of experience in the energy industry – from exploration and production, through generation and distribution, to end-use – where she had successfully integrated SD into business practices. Dr. Sharpe also served on numerous technology and industry association committees, in particular as an international advisor on sustainability issues, and has represented the Canadian energy sector at the Asia-Pacific Economic Cooperation (APEC) Business Forum. She is a known spokesperson nationally and internationally on how to promote an understanding of the culture shift in values society must make to secure a sustainable future. The foundation's first four directors also had considerable relevant experience: David Johnston, president of the University of Waterloo and founding chair of the NRTEE; Ken Ogilvie, director of Pollution Probe; Alain Caillé, vice rector at the Université de Montréal, who had extensive expertise in the research sector, and at various times had been a member on the Board of Directors for NSERC and the Government of Quebec's Fonds québecois sur la nature et les technologies; and finally, Jim Stanford, SDTC's first chair of the Board of Directors, was the CEO and director of Petro-Canada from 1993 to 2000. SDTC has also benefited from the broad experiences of its Member Council, comprising individuals from the non-profit sector, such as David Runnalls, current president and CEO of the IISD, and James Knight, former CEO of the Federation of Canadian Municipalities,[27] as well as private sector members, such as Pierre Alvarez, the president of the Canadian Association of Petroleum Producers.

SDTC's governance framework ensures that there is representation from industry, academia, and the non-government sectors, underscoring the principle behind the creation of the foundation, that partnerships between the various actors were necessary to achieve SD. Though the institutional structure of a foundation to support research and development was not novel, its singular focus on SD was.

GROWING WITH THE BUDGETS, 2000–2007

Unlike other SD institutions in Canada, SDTC does not have a lengthy history. Vicky Sharpe has had a central role in shaping and defining the vision and character of the organisation. The dynamics flowing from the significant increases in SDTC's funding portfolio over its as yet very short life have been extremely influential to its development. From the beginning, Sharpe set out to differentiate her organisation from government. She named the organisation Sustainable Development Technology Canada to eliminate any negative association of foundations as simply being money distributors. Outlining the vision of SDTC as being the impetus for a shift toward SD, she wrote a bold mission statement for the organisation: "SDTC's mission is to act as the primary catalyst in building a SD technology infrastructure in Canada." The use of the word infrastructure here should be noted; it recognises the multiplicity of actors that are involved in bringing a technology to market through the entire supply chain: researchers, product developers, manufacturers, distributors, retailers and end customers. It also emphasises the holistic approach SDTC takes to innovation, and the value it places on partnerships.

Since SDTC selects the most promising emerging technologies with environmental benefits, Sharpe designed a three-phase screening mechanism that would assess the many elements necessary for the technology to be successful. In the first phase, entrepreneurs make an initial application through a simple and straightforward Statement of Interest (SOI), designed to provide SDTC with a good sense of proposed technologies without imposing an arduous application process. These are screened and evaluated by SDTC staff as well as external experts to ensure adherence to selection criteria, which include capabilities in technology, marketing, and business (potential for partnerships and funding). SOIs that meet these criteria are invited to submit a more-detailed proposal, similar to a business plan for the proposed technology. At this second stage, external technical and business experts review the proposals and provide their recommendations to SDTC. Finally, SDTC's Investment Committee and Project Review Committee review the refined shortlist of projects, and present a final list of recommendations to the Board of Directors for review and final approval. These approvals are made in principle, subject to successful contract negotiations. The rigour of this process provides investors the strong quantified validation of proposals that lowers their risk assessment.

Between November 2001 when SDTC got off the ground and November 2002, Sharpe developed and issued two calls for SOIs, outlining the clean air and GHG mitigating criteria for the technologies that would be funded. The response to these first two calls was overwhelming; SDTC received SOIs for over $800 million. This confirmed the need for funding from the clean technology sector, providing legitimacy to SDTC's creation. However, there were delays in running these proposals through the approval process. The Board of Directors, who has final approval over all projects, only had their first meeting in late November 2002. Unfortunately, the government had difficulties confirming and appointing its seven Directors. The first round was thus completed within a year, with a skeletal staff of between two to four individuals, which Sharpe hired by tapping into her network of SD and energy professionals.

In 2003, the Liberals more than tripled SDTC's assets, providing an additional $250 million for climate change technologies and clean air technologies. The funding agreement for this transfer specified that projects were eligible up to no more than 50 percent of the total costs, though the ratio had to be kept to 33 percent for the total portfolio of projects. The timelines were also clear and short: the government required that the funding be totally disbursed in six years, by December 2009. The agreement maintained the focus of the first funding agreement, that 80 percent of the funding was for climate change while 20 percent was for clean air projects. In addition, the government targeted the energy exploration and production and power generation sectors. It specified that over the life of the agreement, SDTC should make available at least $50 million for projects directed to the hydrogen economy and another $50 million for projects related to clean fossil fuels.[28]

Just one year later, the Liberals again substantially increased SDTC's assets in Budget 2004 by $200 million. However, this funding had a completely new focus on clean soil and clean water technologies. As SDTC notes, it represented an important environmental and economic opportunity in Canada; in 2002, the sale of water-related environmental goods generated more than $1.2 billion and environmental-services revenues related to solid-waste management and soil and water remediation generated more than $5 billion.[29] This mandate expansion recognised the abundance of natural resources in Canada, and the need to protect, conserve and efficiently utilise our natural capital. It gave SDTC the opportunity to fund projects that were involved in all of Canada's major economic sectors: energy exploration and production; power generation; energy utilisation; transportation; agriculture; forestry, wood and pulp and paper products, and waste management. Further, the considerable funding increase – from $100 million in 2001 to $550 million in 2004 – was a strong signal by government that SDTC was, in a very short time, meeting expectations.

Notably, SDTC was not marginalised when the Liberals left office. Under Stephen Harper's Conservative Government, a new $500 million funding agreement was signed in September 2007 with SDTC for the demonstration and large-scale development of biofuels technology. This NextGen Biofuels Fund is very different from the previous allocations, under the SD Tech Fund. First, the NextGen Biofuels Fund can provide no more than 40 percent of the total project costs. As Sharpe noted, the substantial sum was required because of the cost of larger demonstration facilities: "Canada has quite a number of very good companies funded from our first fund that now require substantial scale-up to get to market. The private sector is not well equipped or willing to go into this area without assistance. There is a high cap ex (capital expenditure) gap."[30] Second, there has been a significant shift in perspective on the nature of the disbursements: while the SD Tech funding was given to companies as a grant, the NextGen Biofuels money is repayable over a ten year period after the project is completed, and the fund must be completely disbursed by 2017. Third, it also puts an emphasis on commercialisation by requiring the production of plants of commercial size in order to accelerate the market uptake of the technology. This recognition is crucial as the biofuels field is at a capital intensive stage and will require an entirely new infrastructure to support it. Finally, the biofuels issue is one that has a strong potential to become highly politicised, as industry is putting a great deal of pressure on the government to create a favourable environment for the market adoption of this technology.[31] There were however challenges and opportunities in this period where SDTC's budget increased ten-fold in just six years, from $100 million to $1.05 billion, and its mandate was successively broadened.

CHALLENGES

Personnel

Like any successful start-up, SDTC had to incrementally build its organisation from the ground up. The early years saw the organisation double in size annually, from two the first year to nearly forty in 2008. While Sharpe relied on her personal network to attract personnel, the uncertainty surrounding the creation of the foundation and an unwillingness of industry members to move to Ottawa was a challenge in recruiting the expertise required in those first years.

Meeting the increasing demands for information, oversight, and evaluation through the changes in legislation and in successive funding agreements have challenged SDTC's culture and its operating budget. First, it has hired several full-time staff to meet the increasing reporting requirements without a concomitant increase in operating budget from NRCan. SDTC wants to remain a lean organisation by keeping its operating costs below two percent for the total projects, a ratio comparable to the private sector.[32] Yet, the need to hire additional auditing staff because of the increasing reporting and accountability burden, as will be discussed later, jeopardises its lean culture as a greater proportion of employees do not contribute to the core business of the organisation. For instance, there was an 18 month period between 2005 and 2006 where SDTC's offices held more auditors than analysts.[33]

Finances

Like other foundations, SDTC's operational effectiveness is highly dependent on its ability to disburse funding once consortia's proposals are approved. The need for flexibility and speed is important; for some of the companies applying for funding, their projects cannot proceed without the investment, and in some cases delays could mean the death of the firm. It is important to keep in mind that the average time from initial proposal to funding disbursement is nine months, a considerable length of time.

SDTC's first fund, The SD Tech Fund, was renewed through several budgets by lump sums being transferred from the sponsoring departments though Grants and Contributions. These monies were available to SDTC immediately, who then managed the money by placing it in conservative investments. Under the NextGen Biofuels agreement however, SDTC must provide a detailed cash flow forecast to Environment Canada and NRCan, and receives the money on an as-needed basis. This policy decision responds directly to the auditor general's concern that there was an insufficient accountability regime for the transfer payments.[34]

This has had an impact on SDTC's operations: more work is now involved in assessing and documenting the short-term financial needs of the projects which could be approved. In turn, this new acquisition process has slowed the proposal approval process. Together, these two effects have created financial uncertainty for SDTC as it cannot be certain that it has access to the funds as needed. This change in the funding agreement hampers SDTC's flexibility, and has shrunken its arm-length status, placing constraints on its independence.

Shrinking Independence

In addition to operational changes, a number of new reporting and accountability requirements from clauses in its funding agreements with its sponsoring departments, central agencies, and finally, from legislative changes have further constrained SDTC's independence. Yet, from its inception, SDTC has had a more robust suite of accountability measures than some other foundations:[35] the funding agreement, and subsequent agreements, governing the nature of its mandate and scope, annual reports and corporate plans tabled in Parliament, independent third-party evaluations, and audited financial statements.

In SDTC's third funding agreement, Treasury Board Secretariat required Environment Canada and NRCan to add a clause that prevented the Board of Directors from making decisions in a meeting where the majority of directors present were government appointees.[36] SDTC objected to the clause, and raised concern to the Secretariat that as the Board only meets four times a year this new rule would hamper its ability to operate. To get around this problem, the Board passed a by-law that would allow a subsequent vote by the board members who could not be present; this was however still inefficient, and SDTC raised the concern to the Auditor General that it would affect Board governance. In reviewing this clause, the Auditor General determined that "the government's decision to restrict the role of federally appointed directors affected the Board's governance, and interfered with the ability of federally appointed directors to carry out their duties."[37] Ironically, this intervention by the government to prevent the appearance of government control over SDTC operations only created operational difficulties for a well-working system.

The *Federal Accountability Act* (FAA), introduced by the Conservative government in 2006, significantly altered the reporting requirements of foundations. This element of Bill C-2 was a response to the Auditor General's February 2005 Status Report chapter on the accountability of foundations. In it, the auditor general argued that: "Important gaps remain in the external audit regime and in ministerial oversight, two of the three areas examined in this audit. There is no provision for performance audits of foundations that are reported to Parliament. Nor do mechanisms for ministerial oversight adequately provide for the government to make adjustments in foundations where circumstances have changed considerably."[38]

To correct this accountability gap, Part 3 of the FAA extends the application of the *Access to Information Act* to crown corporations and foundations, while Part 5 amends the *Auditor General Act* to include foundations within the auditor general's performance evaluation purview. Both Eliot Phillipson, CEO of the CFI, and Vicky Sharpe had argued strongly against these measures,[39] as they would compromise their ability to protect the confidentiality of their clients, or their highly sensitive intellectual property. They argued that these amendments could have significant negative repercussions on the foundations' operations since entrepreneurs would be unwilling to provide the detail and quality of information in their proposals that was necessary for the screening process if their intellectual property could not be guaranteed to be protected:

While there are some protections under the bill, SDTC would not be able to guarantee confidentiality to our clients, which at this stage in their development of their technologies is of high sensitivity in terms of public disclosure. As I mentioned, the leveraged funding that comes from the private sector is also made on the basis of an opportunity for future profit, which cannot be realized if the information is in the public domain. Therefore, the willingness of entrepreneurs to apply to SDTC and to provide the detail and quality of information that we would need to be able to assess them would be compromised as a result of this uncertainty, thereby significantly limiting SDTC's ability to select the best projects and to obtain leveraging of taxpayers' dollars from the private sector.[40]

The Senate committee reviewing C-2 was sympathetic to these arguments, and included an exemption to allow SDTC's CEO to refuse to disclose certain records containing information relating to applications for funding, eligible projects, or eligible recipients.[41] Unfortunately, the House of Commons rejected this modification, as well as the inclusion of SDTC in section 18.1, which permitted the head of a crown corporation to refuse to disclose a record containing trade secrets or financial, technical, commercial or scientific information that the crown corporation owns and has consistently treated as confidential.

The legislative changes to the FAA and the *Auditor General's Act* have had a direct impact on the foundation model. Previously, the foundations were not directly accountable to Ministers – Parliament could only intervene in the operations of the foundations if they were found to be deviating from their legislated mandated or terms and conditions of their funding agreements.[42] However, as of 2006, the FAA modified the *Auditor General Act* so that the office has the ability to conduct a value for money audit on the operations of SDTC, while changes to the *Access to Information Act* mean the foundation's work is subject to ATI. In addition, the expenses and revenues of SDTC, as well as four other foundations, are now included in the government's annual financial statements. Fortunately, SDTC has not received any problematic ATI requests to date.[43]

The commissioner for the Environment and Sustainable Development favourably reviewed SDTC's activities and processes, and noted that the successive compliance audits and first interim evaluation demonstrated that SDTC was fully compliant with the terms and conditions of its funding agreements. The institutional structure of the foundation is predicated on a fine balance between independence and accountability. As the commissioner noted, "while each government oversight mechanism has an important role, care needs to be taken to ensure that these mechanisms do not create an undue burden or negatively impact SDTC's operations."[44]

OPPORTUNITIES & BREAKTHROUGHS

Making the Cleantech market

SDTC has demonstrated that it is achieving its objective of developing a clean technology infrastructure in Canada: its $342 million in investments has generated a total project value of $1.14 billion. This represents a 2.3:1 leveraging ratio, where the industry portion is 82 percent of the total leveraged funding.[45] This underscores that SDTC is engaging industry to invest in technological development and demonstration, rather than crowding it out, and SDTC's projects involve funding from other levels of government.[46]

One of the indicators that SDTC is filling a niche like no other actor, neither government nor industry, is that it remains the only major player investing at the seed stage in Canada. As tracked by the Cleantech Network, a clean technology industry association, non-SDTC funding at this stage is less $10 million, while SDTC, through creating partnerships, has invested nearly $250 million. Simply put, this phase is too risky for most investors like venture capitalists or angel investors who fund the early and expansion phases, let alone banks and industry who tend to invest at the very end when the product is ready to market. As described earlier, the rigour and thoroughness of SDTC's approval process removes the risk in the viability of the technology, thereby removing the major barriers to investment. Furthermore, SDTC is creating the clean technology infrastructure around which further business is created. The follow-on investing of SDTC-approved projects has consistently risen since 2006; 24 of the companies' projects into which SDTC has invested $70 million have raised $513 million in follow-on funding in the past three years alone. Thus, SDTC's investments have a considerable multiplier effect.

Finally, these technologies are not only producing a tangible economic benefit, but are also reducing GHG emissions and are creating positive environmental benefits. Without SDTC's funding, it is unlikely that many of these technologies' environmental benefits could be certified. As one CEO explained it, companies must have a verifiable and durable claim to emissions reductions, requiring 1000 hours of the technology's use, at full-scale. The cost of this requirement is virtually prohibitive to most companies.[47] SDTC recognises that there is a lag between investment and GHG reduction impact (project completed four years after funding, market penetration 3 years after that), yet the technologies funded so far have the potential to reduce GHG emissions by 135 megatonnes by 2012. However, SDTC is conservative in its projections, discounting this amount by 90 percent to account for possible technical or market failure; their final projected impact is thus 13.5 megatonnes. However, SDTC's real impact cannot be limited to GHG mitigation, as 86 percent of projects produce two or more of SDTC's four types of environmental benefits – GHG reduction, clean air, clean water and clean soil. SDTC's holistic approach, by focusing on SD, demonstrated that the benefits of most clean technologies cannot be siloed.

Timing

Being an early player in bridging SD and the market has provided an opportunity for SDTC to transform business's valuation of sustainability.[48] While discussions with large firms on environmental return have gained no traction, showing the positive impact on revenues is creating an interest in environmental technologies and will lead to a change in perception of these as investment opportunities. This change is evident, even over the short period of SDTC's existence. Several sustainability market indicators have emerged, such as the CleanTech Venture Index, which has grown to an annualised $1.4 billion and whose 30% rate of return is much healthier than the average 19% of other markets.[49] Further, the first report of the recently established Council of Canadian Academies on the state of science and technology in Canada rated SDTC's support for commercialisation favourably in comparison to other organisations, in particular that of the commercial financing sector.[50]

Perception of SDTC as government

Overcoming the perception and prejudices of the business community toward SDTC was one of its early breakthroughs. As she was establishing the organisation, Sharpe needed to make connections to the investment business community, who were sceptical of government involvement in the market. Sharpe needed to convince potential investors, particularly venture capital (VC) firms as they are the most common source of funding for emerging technologies, that SDTC did to not impinge on their investment opportunities, but rather operated upstream from their activities, at the seed stage of investment. SDTC's responsiveness to the market was questioned; would they be influenced by political considerations? While SDTC does try to provide funding across regions, the project proposals are the key factor. Rather than other considerations SDTC must pick winners in order to protect their investment, attract a consortium of players across the innovation chain and, most importantly, successfully leverage funding.

Part of this perception was fuelled by location. Located in Ottawa, Sharpe set up the office at 230 Queen St. nearby other arm's length organizations, such as the CFI and Export Development Canada, to counter the perception that SDTC was just another government organisation. In order to present SDTC as an independent organisation and communicate its objectives, Sharpe met with representatives of all the VC firms. Publicly, Sharpe has portrayed SDTC as a private sector-like institution; in her testimony to the committee on Bill C-2, the *Federal Accountability Act* (FAA) she described SDTC by saying "we operate very much like an early stage venture capital company, providing made-in-Canada solutions to real world industry needs."[51] This language and value system of SDTC were essential in convincing VCs that SDTC was a like-minded institution, willing and able to conduct business on private sector terms. SDTC is thus acting as a sustainability broker, an agent that can successfully engage business leaders on sustainability by using their language to increase relevancy to individual business' values and objectives.[52]

SDTC has also earned its results-oriented reputation by the nimbleness with which it conducts business: there has been a consistently high demand for its services, receiving a total of 1,497 statements of interest in 13 rounds, representing interest from some 4400 different entities. All told, for an organisation whose staff total has never been larger than forty people, this is a considerable achievement.

SDTC's emphasis on its independence from government and its hard-earned reputation has been the vital ingredients in securing trust from the business community. As Minister Goodale testified, it's the individuals that make the foundation: "Anybody who has encountered the Canada Foundation for Innovation will tell you that they are fiercely independent, and that really flows from the calibre of the individuals, people of impeccable credentials that can be counted on to exercise sound but independent judgement. They are not likely to be influenced by the governmental process."[53] The recent interim evaluation suggests that applicants who submit funding proposals are very satisfied with the quality of service from SDTC: "they described SDTC staff as extraordinary and praised them as helpful, technically competent, and easy to work with."[54] Without the business community's trust and perception of SDTC as partner, the organisation could not have achieved its objectives: investment would be sparse, and it would be difficult to attract businesses to partner with the companies seeking SDTC funding.

Capacity building

SDTC has played a very important capacity building role for small-to-medium enterprises (SMEs) in their attempts to commercialise their technologies. As SMEs drive innovation in the clean technology marketplace, but often lack vital experience, SMEs are the appropriate target for SDTC's efforts; this is further evidenced by the fact that 90% of SDTC-funded projects are SME-led.[55] First, SDTC provides assistance to firms building a business plan for their technology, which is a time-intensive process. For example, for those whose SOI is accepted, SDTC will visit applicant-consortia sites to meet with those involved in the projects. If applicants' more detailed proposal is not accepted, SDTC provides considerable support to firms in refining them by giving detailed explanations on the strengths and weaknesses of their business plan and how it might be improved. For this purpose, SDTC designs and hosts "Entrepreneurial Excellence" workshops, which provide forums for mentorship and coaching, and allows entrepreneurs to benefit from hearing other applicants' experience. So far, these workshops have yielded positive outcomes: of the 28 applicants who attended, 14 have gone on to receive funding from SDTC with their revised and improved proposal.[56]

SDTC also provides significant value-added by helping firms to identify consortia members, from industry, government or academia. In this sense, SDTC is building the SD infrastructure by creating partnerships that may otherwise never materialise. As one CEO of an advanced materials company noted, SDTC plays an invaluable role in facilitating the path to market for the technologies: "Going through the SDTC process improved our ability to focus on what was needed to be done to take our technology to market. We were able to develop a clear, solid value proposition that would help us secure future funding from the private sector for full commercialisation."[57]

Ralph Goodale explained to members of Parliament that this is a completely unique role within the innovation system, one that no other government department or program achieves: "none of them orient their funding to private-sector-centred partnerships."[58] SDTC's process is value added to the entities involved, as it sharpens their knowledge and understanding of the market, increases their ability to identify both the economic and environmental strengths of SD projects, and finally, defines the investment potential that their clean technologies ultimately represent to venture capital financiers. It is also building the capacity of the innovation system for these technologies: since inception, SDTC has provided funding to projects which had previously received funding from other government sources, including 24 IRAP programs and 15 NRCan/CANMET projects. The unique assistance at the various phases of the technologies' path to market demonstrates the complementarity between the government's innovation instruments, and underscores SDTC's important niche within that system.

Advancing SD Policy

Though not part of SDTC's legislated mandate, SDTC has become, through its engagement with industry, academia, and government, a substantial knowledge broker of SD. The Statements of Interest SDTC receives gives the organisation a broad and in-depth view on the state of clean-technology development in Canada, the market's needs, players, and potential direction.

One of the most important tools developed by SDTC is their *SD Business Case*, reports which profile specific economic sectors (e.g. commercial buildings or biofuels). They combine data, stakeholder input and industry knowledge to provide an overview of the investment potential in that sector and outline a vision for Canada's future potential. SDTC gathers this stakeholder input through meetings and interviews, and releases the information through numerous town-hall-like sessions across the country. These reports represent a wealth of knowledge on a particular sector, and are shared publicly through SDTC's website.

SDTC is also engaging in policy development more explicitly by partnering with other SD institutions. Using its "bottom-up" knowledge of the market, SDTC is partnering with NRTEE in order to provide advice to the government on how it can encourage the coming to market and dissemination of environmental technologies. This partnership is significant because it shows the synergy of SD institutions; SDTC will have a channel for communication of policy advice at politically senior levels, while NRTEE gains specialised knowledge and insight into a key SD issue. There are also important feedback loops between SDTC and the leading SD policy and advocacy institutions in Canada: Vicky Sharpe is on the Board of Directors of the IISD, while David Johnston, the founding director of NRTEE as well as David Runnalls, IISD's longest serving president, are respectively on SDTC's Board of Directors and Member Council. These cross-appointments provide the opportunity for SDTC to play a role in shaping SD policy by being plugged into, as well as contributing to, the highest level discussions on SD in Canada and internationally.

Finally, SDTC is advancing SD by being a model for emulation. Within Canada, SDTC has played an important leadership role by providing guidance and advice to provincial levels of government interested in developing a funding instrument for environmental technologies. So far, SDTC has received requests from three provinces, Ontario, Alberta and British Columbia, and has expended considerable effort to guide them through the institutional development process. In fact, the senior vice president, Maria Aubrey, spent three days with B.C. Government officials to design their recently announced Innovative Clean Energy Fund.[59]

This influence is not limited to the national level. As an instrument of public policy, SDTC has served as a model for both the Australian and Norwegian governments. Australia announced the creation of the Low Emissions Technology Demonstration Fund (LETDF) in their 2004 Energy White Paper, *Securing Australia's Energy Future*. Similarly to SDTC, the $500 million allotted to the LETDF has the objective of supporting industry-led projects to demonstrate the commercial viability of new energy technologies with low greenhouse gas emissions. However, the independence of the LEFTD seems to have been curtailed under the Labour party, who won the November 2007 election. Upon establishment in 2005, the LEFTD was administered by a third party, AusIndustry, the Australian Government's business program delivery division in the Department of Industry, Tourism and Resources, but was transferred to the Department of Energy, Resources and Tourism as of 1 July 2008. Only five projects were funded in round one, totalling AU$285 million. It is unclear what will be done with the remaining funding as it appears that the fund has been shut down: the Department's website notes that "Round One of the LETDF closed on 31 March 2006, no further funding rounds will be held."[60] The Australian experience underscores the importance of SD focused-organisations' institutionalisation through legislation in order to prevent its activities from being curtailed by political whim before the objectives are achieved.

THE ROAD AHEAD

SDTC has been successful in achieving an impressive leveraging ratio of 2.3:1. However, the source of funding is nearly exhausted. The SD TechFund's clean air and climate change portion of the $550 million is almost entirely committed, and the Board of Directors is now actively focused on securing its renewal. SDTC has received far fewer proposals for clean water and soil, and are now more actively targeting this market segment. The impressive follow-on funding from some of SDTC's completed projects provides evidence that SDTC is an effective stepping stone on the path to commercialisation. Furthermore, this demonstrates that the SD infrastructure in Canada is well on its way.

CEO Vicky Sharpe sees future opportunity in several areas lightly or not yet funded by SDTC: developing a hydrogen economy; biorefineries, which would dovetail well with the NextGen Biofuels Fund; and finally, technologies related to agricultural practices. Despite the growth in the market's interest in SD as indicated by the growth in the CleanTech Venture Index, Sharpe sees SDTC as remaining the only major player in the next years as the markets recede to safer areas.[61] Certainly, recent data suggesting that the Canadian venture capital investment of just over $1 billion in the first three quarters of 2008 is 33% lower than last year.[62] The souring climate of investment emphasises the importance of having an independent organisation focused on this issue as well as a committed pool of funds, independent of governmental surplus or budget-allocation decisions, one that may become a relatively more important source of funding for emerging technologies. Creating an institution that is arm's length, and free from political interference permits the continuity and long-term timeframes which are vital to both SD and innovation. SDTC's funding may provide the market push for environmental technologies, but its de-risking of the technology through the screening mechanism and environmental benefit calculation provides the market pull. Furthermore, by focusing on the development of partnerships between the various actors in the innovation system, SDTC is creating the infrastructure necessary for the maturation of the SD technologies market in Canada. The foundation model has proven itself vital in providing SDTC the nimbleness and flexibility to achieve its objectives. However, the institution's success has been primarily determined by the direction and vision provided by SDTC's executive team and Board of Directors.

NOTES

1 Sustainable Development Technology Canada (SDTC), *Clear Results, Clear Thinking: 2005 Annual Report* (2005), 36.

2 Commissioner for Environment and Sustainable Development (CESD), "Managing the Federal Approach to Climate Change," *Report of the Commissioner for Environment and Sustainable Development to the House of Commons* (2006), 8.

3 Technology Issue Table, *Enhancing Technology Innovation for Mitigating Greenhouse Gas Emissions* (Prepared for the Government of Canada, 1999), 17.

4 Ibid.

5 Ibid.,19.

6 Wayne Richardson, personal interview, 21 August 2007.

7 "Environmental S&T gets $500 million boost in budget," Research Money 14.3 (2000). McGuinty later became a member of Parliament.

8 Leslie Pal, *Beyond Policy Analysis: Public Issues Management in Turbulent Times* (Thomson and Nelson, 2006), 82.

9 Peter Aucoin, *Accountability and Coordination with Independent Foundations: A Canadian Case of Autonomization of the State.* Paper presented to "Autonomization of the State" Workshop (Stanford University, 1 April 2005).

10 House of Commons Standing Committee on Aboriginal Affairs, Northern Development and Natural Resources (AANR) *Minutes of Proceedings* (37th Parliament, 1st session, issue no.3, 13 March 2001).

11 Ibid.

12 Organisation for Economic Cooperation and Development (OECD), "STI Review #25," *Issue on Sustainable Development* (2000), 13.

13 Ken Ogilvie, personal interview, August 2007; Vicky Sharpe, personal interview, 23 August 2007.

14 Sharpe, 2007.

15 Ibid.

16 Department of Finance, *The Budget Plan 2000* (28 February 2000), 117.

17 Mary Preville, personal interview, 24 August 2007. Interestingly, this approach was similar in the creation of the Council of Canadian Academies (CCA). While created in April 2002 under the *Canada Corporations Act*, the CCA did not receive its $30 million in funding untilt the 2005 Budget. See the CCA's website, "Background." Available at: www.scienceadvice.ca/background.html.

18 AANR, *Minutes of Proceedings* (37th Parliament, 1st session, issue no.9, 15 May 2001).

19 Ibid.

20 Ibid.

21 Standing Senate Committee on Energy, the Environment and Natural Resources (EENR), *Minutes of Proceedings* (37th Parliament, 1st session, issue no.10, 29 May 2001).

22 Ibid.

23 KPMG, *Evaluation of Foundations* (Prepared for Treasury Board Secretariat, 2007).

24 Department of Finance, *Accountability of Foundations Backgrounder* (2005). Available at: www.fin.gc.ca/toce/2005/accfound-e.html.

25 Auditor General of Canada, "Accountability of Foundations" *Status Report* (2005), 1.

26 Natural Resources Canada, *Backgrounder: Canada Foundation for Sustainable Development Technology* (July 2001).

27 James Knight was an important member as the Federation of Canadian Municipalities manages the Green Municipal Fund; a $550 million endowment which provides loans and grants to municipal governments for the development of sustainability plans, to feasibility studies and field tests, or to capital projects related to brownfields, energy, transportation, waste, water.

28 Robinson Research, *Evaluation of Sustainable Development Technology Canada: Interim Evaluation Report* (2006).

29 SDTC, "Soil and Water." Available at: www.sdtc.ca; accessed September 2008.

30 Sharpe quoted in Research Money, "New SDTC fund to assist development and demonstration of second-gen biofuels," *Research Money* 21.15, 10 October 2007.

31 For instance, Iogen, Canada's biggest producer of cellulosic ethanol has expressed frustration over government delays in providing assistance. This situation is complicated by the fact that the United States has offered Iogen up to $80 million to construct a plant in Shelley, Idaho, in order to scale-up ethanol production. "New SDTC Fund," *Research Money* (2007).

32 Vicky Sharpe, personal interview, 26 August 2008.

33 Ibid.

34 Department of Finance, *Accountability* (2005).

35 For a good comparison of accountability requirements, see KPMG, *Evaluation of Foundations*, 49.

36 Auditor General of Canada, "Role of Federally Appointed Board Members – SDTC," *Report of the Auditor General of Canada* (November 2006).

37 Ibid.

38 Auditor General of Canada, Status Report (2005).

39 Legislative Committee on Bill C-2, *Minutes of Proceedings* (39th Parliament, 1st session, issue no.15, 2006).

40 Vicky Sharpe, speaking at the Legislative Committee on Bill C-2 (39th Parliament, 1st session, issue no.15, May 30 2006).

41 Peter Niemczak, *Legislative summary Bill C-2: The Federal Accountability Act*, 21 April 2006.

42 KPMG, *Evaluation of Foundations*, 4.

43 Sharpe, 2008.

44 CESD, "Measuring the Federal Approach" (2006).

45 SDTC website. Available at: www.sdtc.ca.

46 Note, the remaining 18% of leveraged funds, or $142 million total, includes funds from provincial and federal levels of government as well as academia.

47 Jeremy Holt, President of NxtGenEmission Control, speaking at SDTC's Annual General Meeting, 18 June 2008.

48 Sharpe, 2008.

49 Vicky Sharpe, speaking at SDTC's 2008 Annual General Meeting, 18 June 2008.

50 Council of Canadian Academies, *The State of Science & Technology in Canada* (2006), 17–28.

51 Legislative Committee on Bill C-2, *Minutes* (2006).

52 Kevin Brady, "Engaging Senior Management on Sustainability," in Glen Toner, ed. *Sustainable Production: Building Canadian Capacity* (University of British Columbia Press, 2006), 202–16.

53 EENR, *Minutes* (2001).

54 Robinson Research, *Evaluation of Sustainable Development Technology Canada* (2006).

55 SDTC, *Capitalizing on Cleantech: 2007 Annual Report* (2007), 19.
56 Sharpe, 2008.
57 SDTC, *Clear Thinking*, 3.
58 Ralph Goodale, Standing Senate Committee on Energy, the Environment and Natural Resources, 15 May 2001.
59 Sharpe, 2008.
60 Government of Australia, Department of Resources, Energy and Tourism. Available at: www.ret.gov.au/energy/energy%20programs/low_emissions_technology_demonstration_fund/Pages/LowEmissionsTechnologyDemonstrationFund.aspx.
61 Sharpe, 2008.
62 "VC investment continues downward trend," *Research Money* 22.18, November 2008.

Contributors

DAVID V. J. BELL is professor emeritus, senior scholar and former dean, Faculty of Environmental Studies, York University. He is currently the chair of Learning for a Sustainable Future (www.lsf-lst.ca); co-chair of EASO – the Education Alliance for a Sustainable Ontario (http://www.sustainable-ontario.org); and co-chair of ESD-Canada (formerly the National Education for Sustainable Development Expert Council [NESDEC]).

SERENA BOUTROS is a recent graduate of the Innovation, Science and Environment stream of the Master of Arts program of the School of Public Policy and Administration at Carleton University. She works for Natural Resources Canada.

FRANÇOIS BREGHA, a senior consultant with over thirty years of experience, specializes in the design of tools and processes to integrate environmental considerations in public policy. He has advised a dozen departments and agencies on their sustainable development strategies and recently completed a major evaluation of the implementation of strategic environmental assessment across the Canadian government.

ANN DALE is a professor with the School of Environment and Sustainability at Royal Roads University. She holds her university's first Canada Research Chair, in sustainable community development. She is a Trudeau Fellow (2004) and a Fellow of the World Academy of Arts and Science and chairs the Canadian Consortium for Sustainable Development Research (CCSDR).

ROGER GIBBINS is president and CEO of the Canada West Foundation, a public policy research group based in Calgary. He is also a faculty professor of political science at the University of Calgary.

LILLIAN HAYWARD is a recent graduate of the Innovation, Science, and Environment stream of the Master of Arts program of the School of Public Policy and Administration at Carleton University. She is currently a policy analyst with the Government of Canada.

JAMES MEADOWCROFT holds a Canada Research Chair in Governance for Sustainable Development and is a professor in the School of Public Policy and Administration and in the Department of Political Science at Carleton University.

ANIQUE MONTAMBAULT is a recent graduate of the Innovation, Science, and Environment stream of the Masters of Arts program of the School of Public Policy and Administration at Carleton University. Her interests in environmental policy have led her to Environment Canada, where she works as a policy analyst.

DAVID RUNNALLS is president of the International Institute for Sustainable Development. He is a member of the board of the Institute of Advanced Studies of the United Nations University and has served for several decades in numerous other capacities to advance the cause of sustainable development in Canada and internationally.

LAURA SMALLWOOD is a recent graduate of the Innovation, Science, and Environment stream of the Masters of Arts program of the School of Public Policy and Administration. Currently, she works in the field of international development as an environment policy analyst.

ANNIKA TAMLYN is a policy advisor at the National Round Table on the Environment and the Economy and a graduate of Dalhousie University's MBA program. Her research interests include the role of business in Canada's sustainable development.

GLEN TONER is professor of public policy in the School of Public Policy and Administration at Carleton University. He is also director of the Carleton Research Unit in Innovation, Science, and Environment and a member of the advisory panel of the commissioner of Environment and Sustainable Development.

DAVID WHEELER is dean of the Faculty of Management, Dalhousie University, Nova Scotia. The Faculty of Management at Dalhousie has a values-based approach to management education and research and graduates women and men who "manage with integrity and get things done" from eight graduate and two undergraduate programs.

MARK WINFIELD is an Assistant Professor of Environmental Studies at York University, and Coordinator of the Joint Master of Environmental Studies/ Bachelor of Laws program offered in conjunction with Osgoode Hall Law School.